T0200859

On Life

On Life

Cells, Genes, and the Evolution of Complexity

FRANKLIN M. HAROLD

OXFORD

UNIVERSITY PRESS

OXFORD
UNIVERSITY PRESS

Oxford University Press is a department of the University of Oxford. It furthers
the University's objective of excellence in research, scholarship, and education
by publishing worldwide. Oxford is a registered trade mark of Oxford University
Press in the UK and certain other countries.

Published in the United States of America by Oxford University Press
198 Madison Avenue, New York, NY 10016, United States of America.

© Oxford University Press 2022

All rights reserved. No part of this publication may be reproduced, stored in
a retrieval system, or transmitted, in any form or by any means, without the
prior permission in writing of Oxford University Press, or as expressly permitted
by law, by license, or under terms agreed with the appropriate reproduction
rights organization. Inquiries concerning reproduction outside the scope of the
above should be sent to the Rights Department, Oxford University Press, at the
address above.

You must not circulate this work in any other form
and you must impose this same condition on any acquirer.

Library of Congress Control Number: 2021945899

ISBN 978–0–19–760454–0

DOI: 10.1093/oso/9780197604540.001.0001

Printed by Sheridan Books, Inc., United States of America

To all who believe, as I do, that the true object of science is neither to accumulate knowledge nor to solve practical problems, but to make the world intelligible.

Contents

Preface

From birth to death we are immersed in an ocean of life, a cornucopia of living things: not only other humans but also animals, plants, insects, and all manner of crawly creatures. Life is so abundant, ubiquitous, and familiar that we seldom give thought to the nature of living things, and what sets them apart from nonliving ones such as stones, clouds, and running water. The object of this small book is to make the phenomenon of life intelligible to readers who are not biologists by profession. What is life, what makes living things tick, how are they related to the world of physics and chemistry, and how did they come to be as we find them? These are the fundamental questions that define biology, and what we have learned deserves to be part of the mental furniture of anyone who aspires to scientific literacy.

Like other portentous words, "Life" has multiple meanings. In everyday speech it refers almost exclusively to human affairs: we are preoccupied with making a living while also living a good life, and some are obsessed with when life begins. The usage here is entirely different, that of the naturalist. Our subject is the parade of forms that share the quality called Life, those living today as well as those known to us only through fossils. We humans hold a place in these ranks, and not a minor one either, but the show is not primarily about us.

In the latter half of the twentieth century the staid science of biology was transformed from a largely descriptive practice centered on natural history into an intensely experimental pursuit, focused on how living things work at the level of cells and molecules. The project has been spectacularly successful, finding answers to questions that could barely be formulated before the Second World War. The nature and general architecture of cells, the mechanism of heredity, and how energy is captured and harnessed have largely been clarified. Microbes have been fully integrated into the life sciences, and evolution is recognized as the overarching principle that makes sense of the diversity of life. Applications to medicine, industry, agriculture, and warfare increasingly rule our lives. At the same time, the gap keeps widening between "us," that is, those who speak Science and take its precepts for granted, and the general public, who understand less and less of what we are up to and are beginning to question our goals and motives. Here, I suspect, is a major cause of the decline in science's standing in our time. I do not presume to bridge that gulf, but do hope to lay down a few steppingstones.

The torrent of discoveries has cast a flood of light on age-old questions that straddle the line between science and philosophy: What defines the living state, how did mindless matter beget purpose and meaning, and how did life arise from the dust of the cosmos when the world was young? What we have learned underscores how extraordinary living things are. Intricate and complicated, they obey all the laws of chemistry and physics, yet the existence of life could never have been predicted from those laws. Life stands squarely within the material world but at the same time stands apart, flaunting its autonomy, purposeful behavior, and in one instance the capacity to reflect on its own nature. Now that we know most of the basics about the way living things work, we need to integrate all that mass of facts into a comprehensible and coherent framework; in a phrase, to make biology intelligible. The question so what is life? is not one for the laboratory scientist: you can't get a grant to study that. But it is an inescapable subject for scientists with a philosophical bent, and over the past two decades I have become obsessed with it.

The trouble is that the volume of biological knowledge is now so vast that it overwhelms the capacity, and the will, of anyone who seeks to grasp large chunks of it whole. Increasingly, the mass of particulars obscures their meaning. I therefore intend to set aside as much of the burden of detail as possible, to extract what seem to me the central principles and to highlight the major questions that biologists ask of nature. This unavoidably entails stepping outside the fields in which I can claim technical expertise, and making personal judgments on matters on which scientists disagree. Science is a journey that remains unfinished, a never-ending conversation in pursuit of understanding. My hope is that what I say here will help readers, and even myself, to better grasp the wonderful and perplexing phenomenon of life.

What do we mean by "understanding"? I use this commonplace term in the sense set forth by the Oxford philosopher Mary Midgley: "Understanding anything is finding order in it. . . . It is simply putting [the object] into a class of things meaningful—noting how its parts relate to it as a whole, and how it itself relates to the larger scene around it."[1] I am not here to present an overview of biology (Ernst Mayr has done that, far better than I could), but to examine the framework of ideas that interpret and explain the facts.

I am a mainstream scientist, steeped in a lifetime of research into the workings of microorganisms, but over the years I have acquired my own glasses through which to view the world. Most of my opinions fall well within the range of conventional thought, with one possible exception. It is fashionable nowadays to minimize the gap between living things and inanimate matter, and to underscore the fact that life is part and parcel of the common physical universe. That statement is assuredly true, but I am even more impressed by the great gulf between things that have life and those that do not. One could say (paraphrasing the geneticist Theodosius Dobzhansky) that nothing in biology makes sense except in the context of chemistry and physics, but everything in biology comes in its own distinctive flavor. Reflection along these lines has engendered an inclination to physiology and complex systems, and a perspective on life drawn from its history. Unlike chemistry and physics which draw on universal laws, biology explores the consequences and ramifications of a singular event, the origin of life. Life as we know it revolves around cells, each of which is an intricate system of myriads of molecules integrated into a unit of form and function. DNA is a database of central importance, but it does not direct cellular operations; those emerge from cell dynamics and are seldom spelled out in the genes. Living things are products, not of design but of the interplay of heredity, variation, and natural selection. Finally, our voluminous knowledge is bracketed by two enduring mysteries: How life began, and how mind arose from matter. I like to describe my attitude as a sort of vitalism, a latter-day or molecular vitalism. Let what I have written here stand as an introduction to an unfashionable point of view.

Acknowledgments

It was my great good fortune to come of age in science at the midpoint of the twentieth century, just as the transformation of biology was gathering speed. Many of the grand masters of molecular, cellular, and evolutionary biology left their mark on my perception of life as a phenomenon of nature, particularly Richard Dawkins, Brian Goodwin, Stephen Jay Gould, Francois Jacob, Lynn Margulis, Ernst Mayr, Peter Mitchell, Jacques Monod, Harold Morowitz, John Maynard Smith, Tracy Sonneborn, Roger Stanier, Gunther Stent, and Carl Woese. I am also much indebted to Ford Doolittle, Nick Lane, William Martin, Nick Money, Denis Noble, Norman Pace, and James Shapiro for conversations and correspondence that made me question what I thought I knew.

Friends and colleagues, some professional scientists and others not, reviewed the manuscript during its gestation: Roy Black, Harold Breen, Stephen Ernst, Donald Heefner, Ronald Merrill, Diana Sheiness, and also three anonymous readers, thank you all. At Oxford University Press I am indebted to my editor, Jeremy Lewis, for letting this book see the light of day; to Bronwyn Geyer, for much help in navigating the shoals of electronic publishing; to Sylvia Canizzaro for meticulous copyediting; and to Saloni Vohra for keeping the project on track despite the pandemic. The diagrams were prepared by Ben Rogers and his team at Vox Illustration.

It's one thing to be conscious of one's forerunners and mentors, and quite another to render due credit. In a book that strives especially to be brief there is no place for the nuanced consideration of particulars, nor for the extensive citations that scientific etiquette demands. Instead, I have chosen a sample of the recent literature to acknowledge books and articles directly pertinent to what I have written here, and to provide a portal for further reading. To those who feel, perhaps quite justly, that their contributions have been insufficiently recognized, I offer my apologies and a reminder that it is the way of science for the best of our productions to be absorbed into the common pool of knowledge, while losing their identity—like raindrops falling into a pond.

I like to think of this book as summing up a lifetime's engagement with science, and one last opportunity to pay tribute to some of those who molded my understanding of life and the universe. One that comes to mind is a craggy Australian in a bush-hat, seated on a log by the side of a trail in the Tidbinbilla Nature Reserve near Canberra. He asked where we had been, and when I told him he said, "You have been walking too fast to look at anything."

Thanks, cobber—lesson taken.

About the Author

Frank Harold was born in Germany, grew up in the Middle East, and studied science at The City College of New York, the University of California at Berkeley, and the California Institute of Technology. His professional career spans forty years of research and teaching, mostly in Colorado. He is presently Professor Emeritus of biochemistry at Colorado State University and Affiliate Professor of microbiology at the University of Washington. Dr. Harold's research interests centered on the physiology, energetics, and morphogenesis of microorganisms, and widened to include life and its evolution. He is also a keen traveler, hiker, and lifelong student of history. Now retired, he remains engaged as a writer, lecturer, and philosopher without license.

On Life

PART I
THE NATURE OF LIVING THINGS

1

Strange Objects

One of the fundamental characteristics common to all living beings without exception [is] that of being objects endowed with a purpose or project, which . . . they exhibit in their structure and carry out through their performances.

—Jacques Monod, *Chance and Necessity*[1]

A Singular State of Matter

Let me begin by stating the obvious: the objects we see all around us fall neatly into two classes, those that are alive and those that are not. Mountains, rocks, clouds, and rivers are "inanimate." Their forms, transformation over time, and eventual fate are determined entirely by forces from outside the objects themselves. The hulking bulk of Mount Rainier is said to brood over my hometown of Seattle, benevolent in some moods and menacing in others; Native Americans traditionally consider it a god. But we know now that Mount Rainier was sculpted by volcanic ructions, by ice and water, and these—rather than any volition of its own—will shape its future. Animals, plants, even microbes are different, quite strikingly so. Living things drive and guide their own activities, whose only discernible purpose is their persistence and reproduction. Their forms are produced by forces of their own making, and are quite faithfully passed from one generation to the next with the aid of an internal program. Inanimate objects are made, living things make themselves.

At the same time, living things are part of the same world of physics and chemistry that rules the clouds and threw up Mount Rainier. That became abundantly clear when, beginning in the 19th century, chemists began to inquire what living things are made of. It turned out that living things are made of chemical substances, lifeless molecules. Their chief elements are carbon, hydrogen, nitrogen, oxygen, phosphorus, and sulfur (CHNOPS), with many other elements present in smaller amounts. Every organism is composed of millions, even billions of molecules, of many hundreds of different kinds. Most of these molecules are found in nature only in the context of living things. Yet the laws that govern the structures and interactions of biological molecules are no different from those that produce inorganic minerals, and most biological molecules can nowadays be synthesized in the laboratory. No "vital force" unique to life has ever been found. So life is chemistry, but chemistry of a very special sort. To borrow an evocative phrase from Stuart Kauffman, it seems that life has explored realms of physics and chemistry that inanimate objects never enter. The more I reflect on this, the more impressed I am by the division of the material world into two classes, things that are alive and things that are not.

As a rule, nature dislikes sharp categories; she prefers her boundaries fuzzy. But in the case of life there are very few ambiguous cases, and most of those vanish on closer inspection. True, one candle lights another, but not by reproduction; the size and shape of each flame are determined by its own wax, not by the donor of the light. Crystals grow themselves and supply seeds for another crop of crystals, but again heredity is not involved. Machines make a subtler instance: intricate and purposeful, machines share many features with living organisms, but even the most sophisticated robot cannot make itself. That day may come and then we shall have to think afresh, but for the present all machines are artifacts of life, accessories to the biological universe but not themselves alive. Freeze-dried bacteria are another intriguing case, because many of them revive when placed in a nutritious medium. They were alive once, and may be alive in the future, but they are not alive now. Still, freeze-dried bacteria underscore a crucial principle to which we shall return: the importance of structure. The only objects that do straddle the line between life and nonlife are viruses. Viruses are obligatory parasites that can only multiply after having infected a suitable host. Simple viruses are "mere" chemicals. They form crystals and some can be made by chemists in the laboratory, yet they grow, multiply, and evolve all too quickly. Besides, their chemical makeup (proteins, nucleic acids, sometimes lipids) assures us that viruses belong to the universe of living things. One can argue that the virion, the virus particle, is not alive, but the consortium of virus and host surely is.

Most readers will probably agree that living and nonliving designate distinct classes of objects, but some will not. Spiritually inclined persons often hold that rivers and mountains also have souls, that everything is alive, and that spirit or mind rather than matter is the essence of the universe. I shall not argue the point, which stems from an altogether different usage of the term "living." Whatever merit there may be in the spiritual take on the world, it seems to miss one of its most remarkable features: that it holds an abundance of those strange objects, material entities that possess "life."

We Know Life When We See It

Life and living are fiendishly difficult to define,[2] but easy enough to recognize. Life is a quality or attribute of objects that draw matter and energy from their surroundings, build and maintain themselves, and reproduce their own kind. It is a prominent phenomenon of nature, just like the tides, earthquakes, and the change of the seasons; not a product of human activities like buildings, nation-states, or poetry. How living things are constructed and how they work are the stuff of modern biology, which we shall sample in what follows. Let me here underscore some aspects that bear directly on the nature of life, and that are necessary elements of biological literacy.

Perhaps the first thing you notice when you begin to explore the universe of living things is its staggering diversity. There are cabbages and there are kings, both living but grossly different. We have towering redwoods, the tiny spiders that live in the crevices of their bark, bacteria and protozoa in the guts of those spiders, and mushrooms all around. What do these have in common, apart from being alive? At first sight nothing at all, but that turns out to be fallacious. When we examine not the forms and workings of the organisms but their chemical makeup, we find a surprising degree of uniformity. All of them are made up of substances of the same kind, such as proteins, nucleic acids, lipids, and a collection of small "metabolites," substances that occur in nature only in the context of living things. This is not to say that all organisms are chemically the same, far from it; but it clearly indicates that all living things are related, members of a huge extended family. We might have guessed that from the obvious fact that we can eat one another (as Darwin did), but the unity of biochemistry underscores a fundamental truth: all life on earth is of one singular kind.

A second, subtler general feature of living things is "organized complexity," visible at every level from chemistry and structure to whole ecosystems. The common term "organisms," in use since the 18th century, implies both organization and complexity. Scholars continue to bicker over just what is meant by complexity, but it is clearly a function of the number of parts and the ways they interact. An airplane is visibly more complex than a bicycle, which in turn surpasses a wheelbarrow. The number of molecules that make up even the simplest organism staggers the mind. A typical individual bacterium is likely to be a short cylinder, not unlike a propane tank in shape but only 2–3 micrometers long and one micrometer in diameter. Far too small to see with the naked eye, we would have to line up 500 of them end to end to reach the thickness of a dime; it would take a thousand billion to fill a thimble. But this minute speck of life holds some 2 to 3 million protein molecules, of several thousand sorts; 20 million molecules of fatty lipids; and some 300 million small molecules and ions. Let's not forget water, the most abundant constituent, some 40 billion molecules. All this in just one tiny cell; an amoeba, a thousand times larger by volume than the bacterium, holds correspondingly more molecules.

Complexity of composition is common in nature; a pinch of mud may rival a cell in the number of components. But the complexity of a cell is different; it has purpose (Box 1.1), and that is what the word "organization" conveys. These are very particular molecules, most of which serve a function (colloquially, a purpose) in the operation of the organism and are assembled into a dynamic interactive system. Many are components of minuscule machines that zip together amino acids to make proteins, transport cargo around the cell, or rotate like a propeller to make it move. Almost all the molecules are arranged in space in a particular pattern that is reproduced in every organism of a given kind. Clearly it takes an awful lot of parts, each one in its proper location, to make a whole, a collective, that operates as a unit to persist and reproduce. It is this organized complexity, its nature and origin, that have come to fascinate me.

The first manifestation of organized complexity was recognized in the mid-19th century: all organisms, large and small, are constructed from basic units that came to be called "cells." Many—in fact, the great majority—are "unicellular": they consist of a single cell. A minority, which includes all the creatures large enough to see with the unaided eye (animals, plants, fungi)

Box 1.1 Biological Order

Order, organization, function, and purpose are slippery terms, but one can hardly think seriously about life without them. I do not wish to enter into the subtleties of definition, but since they recur throughout this book let me be clear about their usage. *Order* refers to regularity and predictability. Patterned wallpaper is a classic example, the solar system is another. Cell architecture is highly reproducible from one generation to the next and represents a higher level of order. *Organization* is a special category of order, defined by the mathematician John von Neumann as order that has *purpose*. The parts of an airliner are arrayed in an orderly manner to the end that the collective, the "plane," can fly. Likewise, the molecular parts of a cell are so arranged as to enable the cell as a whole to persist and reproduce. Most of the molecules and structures that make up a cell have *functions*, in the sense that their loss impairs the performance of the cell as a whole. Gene mutations commonly have that effect. Parts that serve some function can be said to be there for a purpose. The term as used in this book does not imply conscious intent, nor design by any transcendent entity. The only purpose manifest in the existence and operations of cells and organisms is their own persistence and multiplication. If life, or indeed the universe, has any higher purpose it does not leap to the eye.

are "multicellular," aggregates of numerous cells, millions or billions of them. The human brain alone consists of some 100 billion cells. Each cell is itself an organism, a unit of life that makes and reproduces itself. It is invariably enclosed in a surface structure made up of one or several membranes, that keep the cell's interior (its "cytoplasm") separate from the world outside and distinctive in composition and activities. And here is another remarkable feature that should be shouted from the rooftops: cells arise only by reproduction, every cell from a previous cell, and never ever appear de novo out of nonliving matter. Truly, as W. S. Beck said many years ago, "The cell is the microcosm of life for in its origin, nature and continuity resides the entire problem of biology."[3]

As microscopes grew sharper and more powerful, they revealed ever more structure and organization within cells. Initially the term "cell" signified little more than a blob of "protoplasm" bounded by a surface (the "plasma membrane"), with a central dot called the kernel, or "nucleus." By now we recognize two very different kinds of cellular organization: the small and rel-atively simple cells of "prokaryotes," the Bacteria and Archaea, whose true complexity only becomes apparent at the molecular level; and the larger and visibly intricate cells of "eukaryotes," the cells that make up animals, plants, fungi, and protozoa. Eukaryotic cells are endowed with a discrete nucleus, visible chromosomes, a suite of organelles including mitochondria, some-times chloroplasts, a conspicuous network of internal membranes and compartments, and a cytoskeleton. Of the myriads of molecules that make up cells, only some are free to slosh around ("diffuse") in the cytoplasm. Most have a fixed abode in one intracellular structure or another, and their functions demand that they be in their proper place.

Order at all levels—molecular, structural, and functional—is a crucial as-pect of life, and the key to the entire phenomenon. Regularity and predicta-bility are not uncommon in nature (think crystals or the solar system), but organization (purposeful order) is rare. Living things are the only known example, and their degree of order vastly exceeds that of anything else, in-cluding human devices. As Rupert Riedl (1925–2005), a Viennese zool-ogist who pioneered the systematic study of biological order, put it with pardonable hyperbole: "Life is order, pure and simple."[4] Well, it's a peculiar and messy kind of order, shot through with variations, exceptions, and con-voluted contraptions that no self-respecting engineer would tolerate. The reason is that living organisms are products, not of intelligent design but of a long and chancy history.

Life is short, we say, at least shorter than we wish. But the history of life is very long, almost as long as that of the earth itself. Geologists tell us that the earth was formed by the accretion of cosmic dust about 4.6 billion years ago. We do not know when life first appeared, but by 3.7 billion years ago some rocks show indications of having originated in biological processes, and there are even some fossils that almost certainly represent cells. They look exactly like contemporary prokaryotes, the Bacteria and Archaea, and presumably lived in much the same way. The larger and structurally more elaborate eu-karyotic cells (more precisely, fossils generally interpreted as the remains of eukaryotic cells) also go well back in time, but not nearly as far as prokaryotes do: the most ancient of those fossils date to about 1.6 billion years ago. Since eukaryotes, and they alone, gave rise to all the "higher faculties," those some-what ambiguous fossils mark a milestone second in importance only to the origin of life itself. Multicellular organisms make their first appearance later still, a mere 600 million years ago. Over this long span earth's environments changed again and again, most notably from the anoxic state in which life originated to the oxygen-rich atmosphere that prevails today.

This thumbnail sketch of life's history is enormously important, for it supplies a general framework for inquiry into the nature, origin, and even meaning of life. First of all, it is a history, not an instantaneous creation. Taking the findings at face value, it appears that life began nearly 4 billion years ago with small and relatively simple cells of the prokaryotic grade; it took 2 billion years to attain the complexity of eukaryotic cells, and another billion to grow large enough to be seen with the naked eye. For about three quarters of life's history all the life that lived consisted of microorganisms, and they still make up the great majority of organisms in our day. Clearly, all the essential characteristics of life were established at the unicellular level, billions of years ago, and even the first glimmers of memory and purposeful behavior make their appearance among the prokaryotes. Muscles, livers, leaves, embryos, cancer, and immune defense mechanisms all came much later, mind and consciousness later still. It follows that if you seek to understand the nature of life, your proper study is not humankind but cells, specifically microbial cells. This is why this inquiry devotes so much space to organisms seemingly remote from human concerns, especially bacteria, and has less to say regarding higher creatures even though it is these that dominate most of our waking hours. By the time higher organisms appeared on earth, late in the game, the phenomenon of life was deeply rooted and fixed in outline. As Maureen O'Malley puts it, "Philosophy should start with microbes as the entry point into biological reflection and only subsequently focus on larger organisms."[5]

From the perspective of the physical sciences, cells and living things in general are passing strange. They are material objects, but made of molecules found nowhere else arrayed as interactive systems. Living things make and maintain themselves, and act on their own behalf. They make up a large and diverse family of a singular kind. Moreover, in flagrant defiance of the universal tendency of things to fall apart, life has persisted and increased in functional organization over billions of years. How can we explain this unique pattern? The only known answer consistent with the available evidence is evolution by natural causes. Scientists all but unanimously reject the traditional belief in divine guidance, and look to Darwin's ramp of heredity, chance variation, and natural selection to supply a full explanation. The appearance of purposeful design that so impresses the beholder results from natural processes that are in themselves not directed toward any purpose. The role of evolution in the proliferation and elaboration of life is established beyond any reasonable doubt. But mystery continues to linger over the beginning of things: It is simply not clear how functional complexity could have originated in a lifeless universe, to reach the stage where Darwinian mechanisms could take hold. The origin of life has become the most glaring gap in our comprehension of life as a phenomenon of nature. Until we solve it, living things will retain that sense of ineffable strangeness, objects whose very existence seems to call for explanations that reach beyond natural causes.

2

Living Cells, Lifeless Molecules

I see in science one of the greatest creations of the human mind. It is a step comparable to the emergence of a descriptive and argumentative language, or to the invention of writing. It is a step at which our explanatory myths become open to conscious and consistent criticism and at which we are challenged to invent new myths.

—Karl Popper, *Objective Knowledge: An Evolutionary Approach*[1]

The first breakthrough in the quest for a scientific understanding of life was made 150 years ago, with the formulation of the cell theory: All living organisms are made of cells, each cell is itself an organism, and cells never arise de novo but only from preexisting cells. Cells are the atoms of life, and like the atoms of chemistry they are in an important sense "irreducible": when cells are dissociated into their constituent molecules, the distinctive qualities of life are lost. To be sure, those molecules can be isolated, characterized, and even synthesized, and much fundamental insight has been gained thereby. But it is essential to recognize that no molecule, or class of molecules, can be said to be "alive." Life is an "emergent" quality of the collective, the society made up of myriads of molecules working together to perform the many tasks required to keep the society as a whole functioning.

These tasks include drawing matter and energy from the environment and processing them into forms that cells can use; producing all the parts and putting each in its proper place; powering movement and transport across membranes and through space; replicating the instructions that specify the molecular structures; maintaining the cell's integrity and form; and eventually producing two cells where there was one before. The structures, shapes, and interactions of biological molecules can all be described in chemical terms. But when we consider the cellular system as a whole we encounter terms that are not part of the vocabulary of chemistry, notably the concepts of function, information, and evolution by variation and natural selection. Most biological molecules serve a purpose in the operation of the whole; they are elements of a system organized in space. Most biomolecules are never found in nature except in the context of living things, and many show traces of having been shaped by evolution over a long period of time. This is ultimately why biology will never be wholly reduced to chemistry: it is an incorrigibly historical science that revolves around contingent events.

How cells accomplish their many tasks is the province of ponderous textbooks, but I would like to believe that it is possible to make cells intelligible to everyone, just as most of us have a general comprehension of how an automobile works even though we may be shaky on the details. That will be the thrust of the next three chapters.

Cells

Cells are overwhelmingly diverse, scarcely less so than life itself. They range in size from bacteria, a few micrometers in length, to intricately wrought protozoa a hundred times larger, and up to neurons that can reach a meter and more. So what makes a cell? If you ask Google, you will be told that cells are the basic structural and functional units of living organisms, and the smallest units capable of autonomous replication. Each cell contains a collection of genes (its genome) and an assortment of minuscule organelles, organized into an integrated system. True, but most cells contain much more than the minimum, while some lack even a genome. In the end, what defines a cell is its boundary, the plasma membrane. Each cell is a single working chamber, separated from its environment and from other cells by a plasma membrane that controls the flow of materials into the cell and out.

On closer examination the profusion of cell types turned out to consist of endless variations on a very small set of fundamental patterns. By the middle of the 20th century, thanks chiefly to the insights of the microbiologists Roger Stanier and Cornelius van Niel, it had become customary to recognize two categories: prokaryotes and eukaryotes (Figure 2.1). All bacteria are prokaryotes: small and relatively simple cells on the scale of micrometers, whose true complexity is not apparent under the microscope but stands forth at the molecular level. Eukaryotic cells are much larger, roughly a thousand-fold by volume. They are sometimes informally dubbed "complex" cells, because their elaborate organization fairly leaps to the eye. Under the micro-scope, each such cell is seen to be endowed with a nucleus enclosed within a nuclear membrane (hence the name, "true nucleus"); a set of chromosomes that carry the genome and duplicate by mitosis; and a suite of organelles,

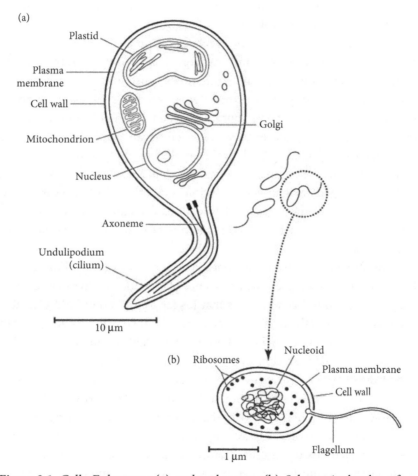

Figure 2.1 Cells: Eukaryotes (a), and prokaryotes (b). Schematic sketches of generalized cells; note the disparity in size and architectural complexity. From Harold 2014, with permission of the University of Chicago Press.

most conspicuously mitochondria and (in photosynthetic organisms) plastids, that serve as the cell's powerhouses. There is a prominent web of internal membranes, and a cytoskeleton that links together the various elements into a working whole. Eukaryotic cells, and only they, make up the large organisms that we are familiar with. We humans are eukaryotes, and so are all other organisms large enough to be seen with the naked eye. There are no forms intermediate between eukaryotes and prokaryotes; here runs a profound cleavage across the entire universe of cellular organisms, to which we shall return many times in this book.[2] Note the omission of viruses, which are not cellular.

Matters became more complicated in the 1970s, when Carl Woese discovered that the class prokaryotes conflates two quite distinct kinds of cells, now termed Bacteria and Archaea.[3] Domain Bacteria houses all the familiar denizens of soil and sea, organisms that cycle nutrients, produce cheese and soy sauce, and sometimes cause disease. Archaea were originally known only from extreme environments such as brine ponds, hot springs, and cow's rumen, but they turned out to be common enough in marine sediments and even in the open ocean. Bacteria and Archaea look much alike, small and relatively simple, but many of their molecular constituents differ in significant ways. We shall see later that Archaea and Bacteria diverged early in cell evolution, some three billion years ago or more. Eukaryotes (now assigned to a third domain, Eukarya) show up much later, roughly 1.6 billion years ago, and represent quite another mode of evolution.

Here is a good place to take note of a striking feature that sets biology apart from the physical sciences: the prevalence of exceptions. Physics and chemistry revolve around general laws that (as far as we know) hold true at all times and places in the universe. By contrast, in biology hardly any generalization can pass without making allowance for qualifications and exceptions. All organisms that we recognize as living are made of cells, one cell or many, but viruses are not—yet viruses are unquestionably part of the biological universe. Prokaryotic cells are tiny, at least the great majority are, but some are large enough to make out with the naked eye and were at first mistaken for protozoa. Eukaryotic cells are usually far larger by volume, but the oceans teem with pico-eukaryotes that feature eukaryotic organization, yet are of bacterial size. As our knowledge grows, the exceptions multiply as well. The protein actin, not long ago regarded as a hallmark of eukaryotes, has now turned up in certain prokaryotes too; and so it goes. We cannot understand the world without making generalizations, but in biology we do so at our peril.

The reason for the variability of living cells is not far to seek. They are products of evolution by random variation and natural selection, and selection has no principles: whatever works in a particular setting, goes. This lends particular significance to those characteristics that do prove to be universal (at least until proven otherwise). The chemical constitution of all cells, prokaryotic and eukaryotic, is universal though with variations. All the world's proteins are constructed from a canonical set of twenty amino acids, a minute sample of all possible structures of this kind. Five bases and two sugars appear in all cellular nucleic acids (but not all viruses). The universal energy currency of all life is a substance called adenosine triphosphate (ATP), and the chief intracellular cation is invariably potassium. Universal facts are less common at the level of cells, but the ubiquity of ribosomes is one such, the division of the cellular world into prokaryotic and eukaryotic grades of organization another. In a world created by evolution universals testify to antiquity, and represent features so deeply embedded in the organization of life that they could not lightly be altered.

The Remarkable Biomolecules

Biochemists are peculiar people, Erwin Chargaff (himself a noted biochemist) once observed. They would take a fine Swiss watch, grind it up in a mortar, and meticulously examine the debris hoping to learn how the watch worked. It's a caricature, of course, but one that highlights an important truth: the molecules of life are not sufficient to explain the phenomenon of life. The difference between a cell and a soup of all the molecules which make up that cell is spatial organization, and we shall have much to say about organization later on. But those molecules are indispensable: one really cannot reflect productively on life and living without some acquaintance with its chemical foundations. For this purpose we can set aside the differences between eukaryotic cells and prokaryotic ones: all cells are composed of the same kinds of molecules, and represent variations on common molecular themes.

Take the large class of molecules called proteins, familiar to everyone because they make up a substantial fraction of the body mass of all organisms and are an essential component of our daily diet. Most of what cells do is done by proteins. They make up the enzymes, the catalysts that govern the rates and specificities of the innumerable chemical reactions that underpin more conspicuous activities. Proteins constitute the engines that make cells move, ferry substances within cell space and transport them across the membranes that mark cellular boundaries. They are major structural elements, the substances that make up much of the cellular fabric, and they carry out many other tasks besides. Proteins are not the only substances essential to life: nucleic acids, lipids, and carbohydrates are no less indispensable, and they will have their due later. But it is the proteins that stand out by virtue of their versatility and prominence in biological activity.

Proteins are polymers, long strings of subunits called amino acids, linked head to tail like beads in a necklace (Figure 2.2A). A typical protein may be made up of some 300 amino acids, but some are shorter and others much longer. All the world's proteins, with some trivial exceptions, are constructed from the universal set of twenty amino acids; some of these are small and simple, but others are quite elaborate molecules. Proteins differ in length, in the proportion of their constituent amino acids, and in the order in which those amino acids are arranged—their "sequence." When we speak of "a protein" (such as hemoglobin, or the muscle protein myosin) we are referring to a population of protein molecules of some particular sequence, all essentially identical and characteristic of the organism from which they came. Beef myosin is not quite identical with horse myosin, but they are similar and obviously members of the same family.

Proteins are linear molecules, but when in their natural state they fold into intricate, three-dimensional knots. Some are globular, others fibrous, each with its own particular shape. Folding takes place even as the protein molecule is being synthesized, and is wholly determined by the sequence of amino acids (provided the environment is permissive). No outside instructions are needed. The capacity to fold up into specific forms is the key to the huge variety of functions that proteins can serve. Each knot holds crannies and crevices that bind substrates and regulatory molecules, and thus facilitate chemical reactions. The active site of an enzyme is one such pocket. There is nothing accidental about these sites: they are reproducible consequences of the way each protein folds, and essential to its function.

Finally, the shapes of protein molecules are dynamic: they "breathe," so to speak. The binding of an enzyme's substrate to its active site subtly alters the local configuration of amino acids, in ways conducive to the reaction. Further shifts in conformation may result from the binding of regulatory

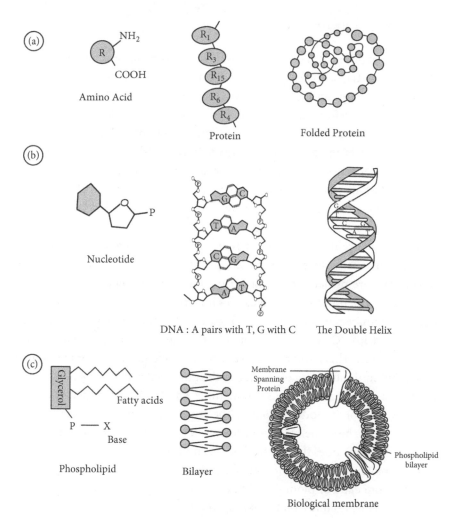

Figure 2.2 A gallery of biomolecules. Left column, the monomers or building blocks; middle column, the polymers; right column, the biologically active forms.

A. Proteins. Proteins are linear chains of amino acids, linked by peptide bonds. All proteins are made from a canonical set of twenty amino acids (R1 to R20), that differ in proportions and in sequence. In working proteins the chains are folded into intricate knots.

B. Nucleic acids. These are linear chains of nucleotides, which again differ in proportions and sequence. DNA is a double helix made up of two antiparallel nucleotide chains, such that the A of one chain bonds with the T of the other, and likewise G bonds with C. In DNA the sugar is deoxyribose, in RNA, commonly single-stranded, the sugar is ribose.

C. Phospholipids. A diverse class of molecules consisting of glycerol phosphate that bears two acyl side-chains. and a nitrogenous base. Phospholipids assemble spontaneously into bilayers in which the acyl groups interdigitate, and the bilayers close up into vesicles.

molecules at other sites on the same molecule. Portions of the structure can thus serve as struts, hinges, platforms, or lids. Sometimes the changes in configuration are large and conspicuous, as in proteins that mediate movements. We should therefore think of proteins as tiny molecular machines (or components of machines made up of several protein molecules) that depend on moving parts and a supply of energy to accomplish some useful task. We shall encounter several examples of molecular machines as we proceed.

Most of what cells do is done by proteins, but not all. In particular, heredity (genes and the expression of the instructions they contain) is famously the business of a second class of large linear polymers, the nucleic acids. To be sure, proteins are also involved in heredity, but the information that genes encode is universally stored and transmitted by a nucleic acid. The monomers in this case are nucleotides, again strung together into chains, and it is the sequence of the nucleotides that carries information, just as do sequences of letters or numbers in our daily lives (Figure 2.2B). There are two great classes of nucleic acids, RNA and DNA, and like proteins these fold up into characteristic shapes. There will be few readers that have not heard of the double helix, the iconic structure of DNA. It is the universal carrier of genetic information in all living cells (viruses sometimes employ RNA), and has become the very symbol of modern biology. The sequence of nucleotides in DNA is what specifies the sequence of amino acids in the corresponding protein, and ultimately determines that protein's form and function. Here we have the essence of biological heredity in a nutshell; all the rest is commentary.

One more class of biomolecules needs to be singled out: lipids, particularly the phospholipids that make up the chemical basis of all biological membranes. Lipids are fatty molecules that shun water and water-soluble molecules, which makes them well suited to the construction of barriers. The basic structure of phospholipids features a fatty molecule at one end and water-loving phosphate at the other. Phospholipids are not large molecules, but they assemble spontaneously into large, closed bubbles or vesicles (not unlike the soap bubbles of our childhood). They form double-layered membranes in which the fatty portions make up the center while the water-loving groups face outward into the watery environment (Figure 2.2C), and such vesicles are the chemical prototypes of cells and cellular compartments. Of course, there is more to living membranes than the barrier. To be useful, a wall must have doors and windows, and these take the form of proteins inserted into the membrane so as to span it from one side to the other. Membranes are sophisticated structures, which will feature often in these pages.

Proteins, nucleic acids, phospholipids, and other biomolecules are all made from workaday atoms—carbon, hydrogen, oxygen, nitrogen, sometimes phosphorus—connected according to universal chemical laws. Why, then, are they never encountered in nature except in the context of living things? There are two reasons, one obvious and almost trivial, the other profound. First, our world teems with bacteria and fungi, all eager to scavenge any stray substance that may serve as energy source or building material. Some even scrounge a living by breaking down such unpalatable fare as petroleum or industrial wastes.; any tasty molecular morsel would not long survive. Organic substances can only persist as parts of a system that continuously protects, produces, and replaces them—to wit, in a living cell.

The second reason organic molecules are basically restricted to the living realm runs deeper. Organic molecules of all sorts are routinely made in laboratories, sometimes under conditions carefully designed to mimic circumstances thought to have prevailed on earth before the advent of microbes. Among those molecules are the simpler amino acids, components of nucleotides, sugars, and lipids, mixed in with many other substances that do not occur in living things. But "prebiotic" procedures never generate specific, functional structures such as proteins or nucleic acids. Some of these can be made in the laboratory with much skill and labor, but they are not the sort of chemistry likely to turn up in the wild. On the contrary, biomolecules are themselves products of the biological universe, and commonly display evidence of evolution by variation and natural selection. That is true a fortiori for the specific sequences of proteins and nucleic acids: ever since the origin of life, particular sequences have been transmitted from one organism to its descendants and shaped to suit a function. Biomolecules belong to chemistry, but their fitness for the functions they serve is a biological attribute, not a chemical one. Here is a door that opens onto one of the grand mysteries, which we will try to pass in Chapter 7.

Viruses

Viruses hold an ambiguous place in the exegesis of life. Originally they were defined operationally as infectious particles small enough to pass through the filters available a century ago; we now recognize them as major components of the biological universe, bridging the gulf between cells and molecules.

Many environments hold more virus particles (virions) than cells, by an order of magnitude or more. Not only do they infect us humans, sometimes with catastrophic results (e.g., smallpox), but also they attack all other kinds of life including Bacteria and Archaea; no one, it seems, is free of viruses. In natural environments virus infection is a major cause of mortality, putting a brake on the proclivity of cellular life to multiply indefinitely. Viruses lack the capacity for metabolism, energy conversions, and protein synthesis; but they multiply all too effectively by commandeering the cellular machinery of whatever cell they infect, usually killing the host in the course of their own multiplication.

Viruses make up an exceedingly diverse universe of their own, ranging in size from nanometers to micrometers. Most are comparable in size to ribosomes, but the recently discovered mimiviruses are so large that they were at first mistaken for bacteria. Some of the smaller ones consist of no more than one molecule of nucleic acid, DNA or RNA, which houses a handful of genes, plus a coat of proteins specified by those genes. They are spherical or filamentous, some can form crystals, and some have been chemically synthesized in the laboratory. Other viruses are quite shapely: bacteriophages, viruses that infect bacteria, take the form of a moon-lander, with a hexagonal head and splayed legs. Some viruses are enclosed in lipid membranes, and mimiviruses boast at least some metabolic capacities (but are still obligatory parasites, on protozoa). At the lower end of the scale, viroids and plasmids are yet more reduced than viruses, consisting solely of infectious nucleic acid. Together, viruses and their cousins span the scale from molecule to cells.

Thanks to the simplicity of their structure and life cycle, viruses (especially the bacteriophages) continue to play a key role in the development of molecular biology. In a nutshell, having made landing on the surface of some susceptible cell, the "phage virion" injects its own DNA into its victim's cytoplasm. There it associates with the cell's DNA, undergoes repeated replication, and assembles into fresh virions, commonly producing about a hundred progeny in an hour or less. The host undergoes lysis, liberating viral offspring into the medium avid to seek out a new host. Many of the major discoveries of molecular biology were first made by studying virus replication, including compelling evidence that genes consist of DNA rather than protein, and that biological forms can arise by self-assembly of their molecular constituents.

Should we consider viruses to be "living"? The answer depends on your definition of life (Chapter 1, note 2). If you stand with Francisco Varela and Lynn Margulis, holding that living things make themselves, then viruses are not alive. Only the consortium of the virus plus its infected host qualifies. If, instead, you stand with John Maynard Smith in putting the capacity for evolution by heredity, variation, and natural selection central, viruses fit the bill. Either way, viruses are clearly a part of the biological universe: they are made of the same substances as cells (nucleic acids, proteins, sometimes lipids), and they make their living by interacting with cells of all kinds. My own preference is to recognize that viruses straddle the line between lifeless molecules and living cells. That should be neither surprising nor alarming: nature likes her boundaries fuzzy. Sharp lines are something we humans read into nature, not something we find there.

3

Life Makes Itself

We have always underestimated cells. Undoubtedly we still do today. But at least we are no longer as naive as we were when I was a graduate student in the 1960s. Then, most of us viewed cells as containing a giant set of second-order reactions: molecules were thought to diffuse freely, randomly colliding with each other. . . . [Today] the entire cell can be viewed as a factory that contains an elaborate network of interlocking assembly lines, each of which is composed of a set of large protein machines.

—Bruce Alberts, "The Cell as a Collection of Protein Machines"[1]

Life is not hard to recognize but fiendishly difficult to define. Among the current definitions, one of my favorites was proposed in the 1970s by two Chilean scientists, Umberto Maturana and Francisco Varela, and promoted with her customary brio by the late Lynn Margulis: Living things are autopoietic systems, systems that make themselves. They produce their own substance, put everything in its proper place, maintain their integrity over time, grow and reproduce their own kind. All the directions come from within. We have long been familiar with these facts and are apt to take them for granted, but the more one reflects on them the more astonishing does the process of living become. The object of this chapter is to give readers a general sense of what the process of living entails.

A Hive of Purposeful Chemistry

Back in the salad days of biochemistry, a standard item of laboratory decor was the wall chart of cellular metabolism, a diagrammatic overview of all the chemical transactions that underpin life. Most young biochemists cut their professional teeth on one or another subset of those reactions (the conversion of sterols to bile acids in the rat, in my case). At first sight it seems that any reaction that can go, does go; but closer examination reveals purpose and order in the welter of chemistry, as summarized in Figure 3.1. First comes a collection of preparatory reactions that funnel nutrients and energy sources into the metabolic web. Next, the core of the web consists of reactions that produce a small number of more-or-less universal metabolites which carry matter, reducing power and energy all across the web. One prominent member of this set is adenosine triphosphate, abbreviated ATP, which serves as the universal energy currency of all life, much as money does in the human economy. Cells expend ATP (by lopping off one or two of the terminal phosphoryl groups) to purchase goods and services, such as biosynthesis or transport. The function of the great highways of metabolism, photosynthesis, and respiration is to replenish the ATP as quickly as it is consumed. How that is done will be the

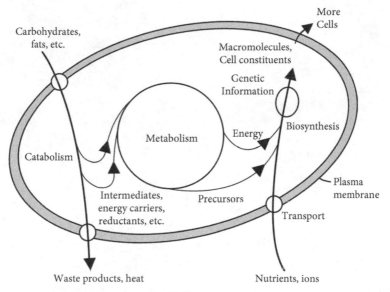

Figure 3.1 The cell as a chemical factory. For discussion, see text. From Harold 2001, with permission of Oxford University Press.

subject of the following section. Next comes a large set of diverse pathways that produce cellular building blocks (amino acids, nucleotides, lipids, and carbohydrates), also coenzymes, signaling molecules, and much else besides. Finally come the processes of growth, the assembly of proteins, nucleic acids, membranes, and walls, each in its correct location.

All this transpires at a furious clip, and in a cramped and crowded space. A cell of the bacterium *Escherichia coli*, of all living things the one most fully understood, takes the form of a short rod about 2 micrometers in length and 1 micrometer in diameter, which holds millions of molecules of diverse substances (Figure 3.2). Under good conditions a cell can duplicate itself every forty minutes, at a cost of about 10 billion molecules of ATP. Since a cell

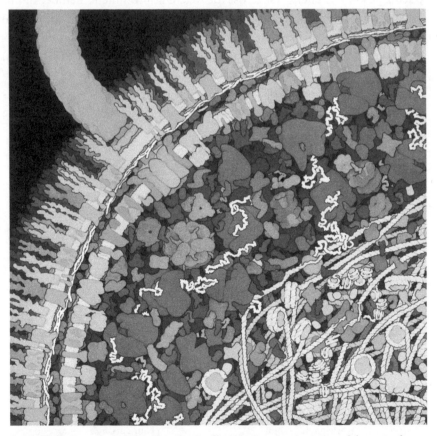

Figure 3.2 A crowded, bustling hive of biochemistry. A vision of the cytoplasm of *E. coli*, with components drawn to scale. Illustration by David Goodsell, The Scripps Research Institute.

holds only about 3 million molecules of ATP at any time, the ATP pool must be regenerated every second![2] Numbers like these give substance to the metaphorical description of a growing cell as a chemical factory.

What confers order on the cacophony of jostling molecules is enzymes, the characteristic catalysts of biology. Catalysts, by definition, accelerate chemical reactions but themselves emerge unchanged; metal ions often have this effect. Enzymes are catalysts with a difference, capable of accelerating selected reactions a billion-fold, sometimes more. The term reaches far back into the history of biochemistry. It comes from the Greek for "leaven," and references the discovery at the end of the 19th century that extracts of yeast devoid of living cells could carry out the conversion of sugar into alcohol. The factors that accomplish this feat could henceforth be investigated apart from cells; in vitro, as we say (in glass, i.e., in a test tube). Enzymes turned out to be proteins, whose isolation is nowadays a fairly routine task (which, in turn, revealed that some enzymes actually consist of RNA rather than protein; we shall have more to say about ribozymes in Chapter 7). The catalytic prowess of enzymes has two important consequences. First, enzymes allow chemical reactions to proceed at ambient temperature, without the heating that chemistry often requires. Second, enzymic catalysis is extremely specific, guiding the flow of matter and energy into particular channels and minimizing the formation of byproducts. Enzymes are made-to-order catalysts selected in evolution for a particular function. That does not mean that they achieve perfection, only that they are good enough to serve the cell's needs.

What is it that makes proteins such potent catalysts? Proteins, as we saw in the preceding chapter, are mobile structures. When the substrate binds to the active site of an enzyme, the protein's conformation is so altered as to bring catalytic groups into close proximity to each other and to the substrate; this facilitates the reaction, followed by release of the products and restoration of the original configuration. Note also that enzyme-catalyzed reactions are intrinsically "vectorial": they have a direction in space relative to the enzyme protein. That is of no consequence so long as the protein is in solution and free to tumble every which way. But when proteins are embedded in membranes, reaching across from one side to the other, then their intrinsic directionality may become the basis for the translocation of a substrate or a chemical group from one side to the other.

Metabolic enzymes sometimes operate as single protein molecules in solution, but many proteins are likely to be members of some large coordinated complex consisting of ten to a hundred individual protein molecules of different kinds. These are veritable machines, the molecular engines that make all cellular operations work. Clever devices of this sort generate ATP, zip together amino acids into a protein, transcribe the information encoded in genes and then translate it, move substances across cellular space, fuse membranes together, or bud off vesicles. A century ago it was acceptable to muse on a mysterious vital force that kept living things active and doing, but no longer. Today we see a cell as a frantically busy construction site, humming with molecular machinery, all driven by chemical reactions that either generate energy or put it to work. Something has been lost in that change of perspective, the sense of enchantment that I recall from my own first forays into the realm of living chemistry. But if life has lost its aura of ineffable mystery, it remains as wondrous as ever that such hives of purposeful chemistry should exist all around us.

Putting Energy to Work

In the inanimate world events follow a predictable course. Water flows downhill, hot bodies cool, paint peels off the wall; things fall apart. Living things are different: they actively seek food, keep themselves warm, and maintain their structure in the face of decay; they grow, multiply, and evolve over time, generating mounting levels of organization. This in no way contradicts the laws of physics, because living things pay the bill by capturing energy from a source in their environment and harnessing it to the performance of work. They are not entirely unique in this regard: every washing machine or automobile does the same. Note, though, that machines are man-made objects, and to that extent themselves part of the biological world. Machines are not alive, and organisms are not machines. But they share important and instructive features, and the metaphor of the machine pervades biology.

There is no mystery about what "energy" and "work" mean in everyday life, but now we need to be a little more rigorous. Work is the easier to grasp. Climbing stairs is a familiar example of physical work, and physicists measure that work by the force required and the distance over which it acts. Pumping up a bicycle tire is work, and so is the accumulation of any substance (e.g., potassium ions) by living cells. Synthesizing a large and complex molecule with a specified structure, such as a protein, is also a kind of work. Broadly speaking, doing anything that goes counter to the spontaneous tendency of natural events is work. So air leaking out of the tire, the spontaneous tendency, requires no work; pumping air in takes work, and can only be ac-complished with an input of "energy." Most of what organisms do qualifies as work, and work requires energy.

So what is energy? Its actual nature is elusive, and would take us too far afield. Let us be content with defining it as the capacity to do work, a quality possessed by matter whenever it is in an unstable state, far from equilibrium. A can full of gasoline is in principle unstable; a match can set it alight, re-leasing a burst of energy in the form of heat and light. But that energy can also be harnessed to do work by making my car run up the hill. That is what living things do: they harness natural sources of energy to do the kinds of work they find useful, including movement, transport, biosynthesis, and maintenance.

The world is full of processes that can potentially serve as sources of energy—a burning log, the sun's heat, lightning, wind, the tides, and so on—but living things rely almost exclusively on two of them: light and the breakdown of preformed organic substances by chemical reaction with at-mospheric oxygen. (This bland assertion omits those organisms that live by geochemical reactions, which will become prominent in Chapter 7.) The phrase to conjure with is "energy coupling": How do living things capture the energy inherent in the oxidation of, say, a lump of sugar (or in a beam of light), and harness it to the performance of work? We all know that an automobile engine carries out a controlled combustion of gasoline, chan-neled by the design of the engine and its accessories so as to make the wheels go 'round. What is the equivalent process in living cells? This is the "cou-pling problem," and by the middle of the 20th century it had become the burning issue in biochemistry. Its resolution by Peter Mitchell in the sixties made a revolution in bioenergetics and forged an unexpected link between the harvesting of energy and the organized structure of living matter. For me personally, the years of engagement with Mitchell and bioenergetics shaped all my subsequent reflections on life, the universe, and all that.

The heart of the matter is that living organisms harvest energy in electrical form.[3] What happens is not unlike what goes on in a power plant, where en-ergy from burning coal or oil is converted into a current of electrons; elec-tricity is portable, and can be connected to many appliances. Likewise, organisms process foodstuff chemically and ultimately feed the products into a device (the respiratory chain) that "burns" them and captures the energy released in the form of a current of charged particles. However, in organisms the current is carried by particles that bear a positive charge, most com-monly protons (hydrogen ions, H^+). Light energy is harvested in a similar manner: light is absorbed by specialized pigments including chlorophyll, and eventually converted into a current of protons (Figure 3.3).

The connection to cell structure stems from the fact that energy capture and transformation take place at the lipid membranes that bound all cells and organelles. The proteins of the respiratory chain (and also the apparatus of light absorption) are built into and across a membrane in such a way that, when sugar is oxidized, protons are pumped from one side to the other. Lipid membranes naturally form closed vesicles that are largely impermeable to protons and other substances, except by way of specific portals. Protons bear a positive charge, and their extrusion generates an electrical potential across

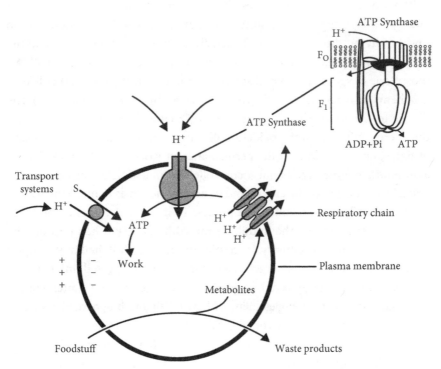

Figure 3.3 Energy coupling by a proton current. The diagram depicts a bacterial cell that lives, as we do, by respiration. The respiratory chain is built into and across the plasma membrane, such that respiration pumps protons from the cell's interior into the external medium and generates an electrical potential, interior negative. Protons flow back into the cell through the ATP synthase, which couples the flux to the synthesis of ATP. Also shown is a transport carrier for substrate S, which links uptake of S to the downhill flow of protons. Insert: The ATP synthase.

the membrane (vesicle interior negative) that tends to pull the protons back whence they came. But they can only respond by flowing through devices cleverly constructed so as to mediate proton flux while harnessing the energy to some useful purpose (for a familiar analogy, think of the turbines that harness the flow of water through a dam to generate electricity). Some cell functions are driven directly by the proton current, particularly transport across membranes, but the most important of the devices that harness the current is the ATP synthase, an intricate membrane-spanning rotary motor that uses the energy to produce ATP (Figure 3.3). That is one of biology's key molecules, the universal energy currency, that cells expend to purchase such services as protein synthesis, transport across membranes, movement, and even the transmission of information.[4]

Energy coupling by means of an ion current is a complicated and counterintuitive way to harvest energy. Why do it in such a roundabout manner? One answer is that a proton current makes a very flexible mechanism: any energy source that can be made to drive a proton pump becomes available for biological work. It also provides a means to accumulate small increments of energy until they add up to a useful amount, like pennies in a piggybank. But I suspect that a larger reason is buried deep in the past. Energy capture by way of ion currents has proven to be one of the universals of biology. All cells, apparently without exception, make use of an ion current to couple energy to work (albeit with numerous variations on this general theme). To the best of my knowledge, no cell lacks the ATP synthase and some of its accessories. This discovery is quite certainly rich in meaning. Ubiquitous processes are probably very ancient, invented early and retained ever since. The universality of ion currents suggests that membranes and chemical reactions organized across them were part and parcel of cellular life from the start. In the Beginning was the Membrane!

Biologists are forever looking both ways, seeking unity while not losing sight of life's diversity. Prokaryotic and eukaryotic cells rely on the same principles but deploy them in different ways. In prokaryotes, energy conservation and transduction is one of the primary functions of the plasma membrane, as sketched in Figure 3.3. Eukaryotes entrust the generation of ATP to specialized intracellular organelles, mitochondria for respiration and plastids for photosynthesis (Figure 2.1), which leaves the plasma membrane free to concentrate on the uptake of nutrients and communication with the outside world. And yet, unity has the last word. Both mitochondria and plastids operate in the prokaryotic manner, and are now known to be the descendants of bacteria that were once free-living but took up residence in in eukaryotic cytoplasm more than a billion years ago. Here is another door that opens unto mystery, this time the origin of the eukaryotic cell, and we'll push it open in Chapter 6.

Heredity

One biological universal that everyone is familiar with is that "Like begets like." Whether we speak of microbes, mice, or men, offspring closely resemble their parents in all fundamental respects. They are not identical, but they are invariably of the same kind and share innumerable traits both physical and mental. We take this for granted, and reckon biological heredity—the accurate self-reproduction of a pattern—to be the feature that most reliably distinguishes living things from inanimate ones.

The discovery of how heredity works is probably the outstanding achievement of 20th-century biology. By now the subject is taught in school, and most readers will be generally familiar with the essentials; let a brief and simplified recapitulation suffice here. As outlined in an earlier section, most of what cells do is done by proteins; they make up the enzymes, transport carriers, scaffolding, and structures that constitute the machinery of life. Recall that proteins are linear molecules, composed of amino acids strung together head to tail in chains 300 to 1,000 units long. All proteins are constructed from a standard set of 20 amino acids, which are among the universals of biology, but the number of amino acids and their order (their sequence) varies from one protein to another. Transmission of a trait or character (e.g., the ability to use lactose) from parent to offspring requires instructing the latter to make proteins of the correct sequence. The proteins themselves are not usually passed to the next generation, rather it is the capacity to make those proteins.

The instructions for making proteins are stored in another linear molecule, this one a nucleic acid called deoxyribonucleic acid, or DNA, which is the genetic material of all living cells. DNA consists of two complementary nucleic acid chains wound around each other to make a double helix. When cells reproduce, DNA is duplicated precisely; one copy remains with the mother cell, the other goes to the daughter. The information that specifies the amino acid sequences of proteins is stored in DNA in the form of sequences of nucleotides, again like letters in a word. To a very first approximation, it helps to conceive of a gene as the length of DNA that specifies one particular protein, which in turn corresponds to one simple trait (such as the ability to use lactose).

In itself, DNA does nothing; it is a repository of sequence information, which only becomes meaningful when it is "expressed" by an exceedingly intricate and sophisticated procedure (Figure 3.4). The DNA sequence (e.g., ABDC) is first "transcribed" into a more portable sequence, another nucleic acid called messenger RNA (e.g., abdc). This is then "translated" into a different language, the sequence of amino acids in a protein. Each amino acid is specified by a particular triplet of nucleotides in the messenger RNA; the table of correspondences is known as the genetic code, which is the same for all organisms on earth. A triplet of RNA nucleotides binds to its cognate amino acid (in an "activated" form), and the complex feeds into a slot of a specialized machine called a ribosome. It is ribosomes that zip amino acids, one by one, onto the growing chain of a fresh protein, in a sequence ultimately specified by the sequence of bases in DNA. Ribosomes, again, are in principle universal: all cells make proteins with the aid of ribosomes, which operate by a universal mechanism but differ in detail among the major kinds of cells. As newly made proteins emerge, they fold spontaneously into the configuration required for their function in the cellular economy. We shall return to ribosomes many times in what follows.

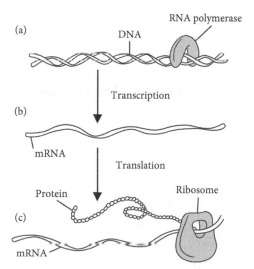

Figure 3.4 This we know: DNA makes RNA makes protein. Adapted from Harold 2001, with permission of Oxford University Press.

DNA replication, transcription, and translation are all highly accurate, but occasional mistakes do happen. An error in DNA sequence, a "mutation" is likely to be transmitted to the next generation; it may cause a change in the amino acid sequence of the protein that gene specifies and quite possibly alter or abolish its function. Mutations are an important source of biological variation, and therefore serve as one of the raw materials of evolution.

So is this all the story of biological heredity? Certainly not. First of all, at the cellular level the mechanics of heredity differ considerably from one kind of organism to another, and especially between prokaryotes and eukaryotes. Eukaryotic cells have chromosomes, nuclear membranes, genes chock-full of insertions, and a long roster of regulatory factors; these are not found in prokaryotes, or only in much simpler forms. Even if we focus on just the molecular level, genes commonly specify a number of features in addition to protein sequences, and not all that genes specify hinges on the transcription of nucleotide sequences. Nowadays there is much interest in "epigenetic inheritance," the transmission of traits as a result of superficial modifications in DNA structure that do not involve alteration of the nucleotide sequence. At the level of the whole cell, form and spatial architecture are quite faithfully transmitted down the generations but are not spelled out in the genes, and depend on cellular mechanisms that are still not fully understood. All the same, the mantra that "DNA makes RNA makes proteins," and also the "Central Dogma," which states that information is carried from nucleic acids to proteins but never in the reverse, encapsulate principles of transcendent importance. They remain indelible elements of the elusive secret of life.

Integrated by Information

The chemical perspective puts the spotlight on the molecular parts of the cells, their chemical makeup. Important as these are, they tell but half the story. A living cell is a functional whole, an intricately organized system of parts working together. And just as we had to invoke a nonmaterial agency, energy, to understand how cells do work, so we must invoke another to make sense of their organization: "information." Francois Jacob, one of the giants of early molecular biology, put it in a nutshell years ago: Energy is the power to do, information is the power to direct what is done.

Consider what we mean when we say that a gene specifies the sequence of amino acids in its corresponding protein. What is it that passes from gene to protein? Not matter—not a single atom of the gene's DNA ends up in the protein. Not energy, either. Translation of one sequence into another is not a spontaneous process, it's a kind of work and requires energy input. But that energy does not come from the DNA, which remains undiminished to serve over and over again. It comes from the recipient, the cell. What passes is something intangible, a reading of the DNA's pattern of order, and that is what we call "information." The information content of DNA derives from the enormous number of ways in which its nucleotides could be arranged (10^{540} options for the gene that specifies a protein 300 amino acids long, an inconceivably vast number). So any gene carries an enormous amount of specified order, which is transmitted to its protein product.

Sequence information is the mode of information that gets the headlines, but it is by no means the only kind significant for organisms. Living things fairly bristle with devices that register some aspect of the environment, both external and internal, and transmit that data to an effector that can act on it. Take but a single example from the bacterial world. Arginine is one of the canonical twenty amino acids, found in every protein. Bacteria usually synthesize their own arginine, but if the amino acid happens to become available in the surroundings (courtesy of a passing cow, perhaps), they take advantage of the windfall. They respond in two ways: the first enzyme in the pathway dedicated to arginine biosynthesis is inhibited, and if arginine is sufficiently plentiful the production of the entire pathway is shut down as well. Obviously, the appearance of arginine in the surroundings constitutes information that is economically useful to the organism. Again, what is it that passes from arginine in the medium (as measured by an appropriate receptor protein) to the cellular effectors? Neither matter nor energy but something intangible, an assessment of an environmental parameter such as the concentration of a desirable substance. What passes is again information, meaningful information.

These two modes of information transfer, based respectively on macromolecule sequences and on signals that report the state of the environment, are of incalculable importance in directing what work an organism can perform. Genetic information is the core mechanism that defines the identity of every cell and organism; signals underlie all its responses to changing circumstances. The two modes are not identical. Genetic messages instruct, signals merely flip switches. Also, information in genes passes unidirectionally from the nucleic acid level to that of proteins, never in reverse. By contrast, signals pass both ways, from the genome to cellular effectors and from receptors to the genome and to effectors. But they share important features that allow us to recognize both as instances of information transfer. Both kinds convey meaningful readings of how matter is organized, both elicit useful responses, and in both cases the requisite work is done not by the sender (as in a radio message) but by the recipient's own organized machinery.

Information is an abstract and slippery concept, more readily defined by what it does than by what it "is." The term is also used in more senses than one. What information means to biologists is not quite the same as what communications engineers had in mind when they defined the concept seventy years ago. In biology we are not concerned with information as such, only with information that carries meaning for a cell or organism (I am greatly indebted here to a pithy article on meaningful information by Anthony Reading,[5] which did much to clarify my own understanding of the concept). Even within the realm of biology, there are subtle distinctions. The explicit meaning of a sequence of nucleotides would be the sequence of amino acids in the cognate protein. Implicit meanings range more widely: they include the protein's three-dimensional architecture, it's functions in the cell's economy, and sometimes ripples that extend beyond the organism itself. Information is as much in the mind as in the message, and that may be why it is not as prominent in biological discourse as are cells, molecules, or even energy.

But make no mistake about it: energy and information, abstract as they are, are the two causes of change in the world. Energy transactions pervade the universe, living and inanimate both; energy is even interconvertible with matter. By contrast, information is strictly a biological agency, found in nature only in living things and in devices fabricated by them. It is one of the characteristics that define the living state: we are all dynamic systems, animated and sustained by flows of matter, energy, and information.

4

Putting the Cell in Order

Our task now is to resynthesize biology; put the organism back into its environment; connect it again to its evolutionary past; and let us feel that complex flow that is organism, evolution and environment united.

—Carl Woese, "A New Biology for a New Century"[1]

Among the books that shaped our intellectual development, scientists of my generation commonly list a small slim volume titled *What Is Life?*, by Erwin Schroedinger. One of the leading physicists of the 20th century, Schroedinger fled his native Austria when the Nazis marched in and found a haven at University College in Dublin. His contract required him to present a series of public lectures, which were published in 1944 to great acclaim.[2] Schroedinger sketched a physicist's view of life, with the focus on the nature of genes and the thermodynamics of biological order. Indeed, he did more than that: he laid out the agenda for a modern biology housed under the umbrella of the physical sciences. And his timing was perfect. With the war winding to a close, hundreds of young scientists were eager to put their talents to nobler use and took up the challenge of an ancient riddle restated. Among molecular scientists Schroedinger is likely to be remembered today for his prescient insight that the genetic material must have the structure of an "aperiodic crystal." No less seminal were his musings on biological order, which laid the foundations for the contemporary approach to this vital but elusive aspect of all living things.

A Singular Kind of Order

First state the obvious: Living things, from animals and plants to single cells, are extremely ordered entities. When I tell you that I saw a great blue heron down by the creek, that label calls up a host of regular, predictable features of form, physiology, molecular structures, and behavior. Moreover, that order has function, or purpose: to catch fish, persist, mate, and raise a new generation of herons. Biological order is not precise; it comes with flaws and numerous variations, but even so it greatly exceeds the orderliness of all other natural objects.[3] The aim of this chapter is to consider how biological order is put in place and how it is perpetuated in the context of single cells. But before we go there, we must take a closer look at the special mode of order that distinguishes living matter.

Order, in the sense of regularity and predictability, is common in nature; think of the solar system and all the regularities linked to it, from the checkerboard of nights and days to the tides and the seasons. Organization, order that has purpose, is very rare. Life is the prime example, but there is another class of objects that are purposefully ordered: our own machines, from bicycles to computers. Machines consist of parts arranged in space to form a collective entity that draws in energy, sometimes matter also, and performs useful work. The analogy to living things leaps to the eye, and the metaphor of the machine (or mechanism) has pervaded biology for centuries. Many cellular organelles are quite plainly mechanisms— ribosomes, for example, or flagella. But what about the whole cell: are cells mechanisms of sorts, more sophisticated than the sort we build but fundamentally of the same nature? The answer hinges on what we take to be essential. Living things are concrete, material objects; if they possess a vital principle or spiritual essence unique to living things, it has never been found. To that extent they can be labeled mechanisms, but this designation does not illuminate anything of significance. Organisms make themselves, repair themselves, and perpetuate themselves; would that my automobile could do so well! Machines are built for a purpose defined by their designer, living things are products of evolution and have no discernible purpose or function beyond their own existence. Moreover, since all mechanisms are products of living things they are subordinate to life, not models for it.

In a world ruled by the Second Law, which mandates that all complex objects must decay over time, order is an anomaly and organization doubly so. It can only be maintained by doing work to repair or replace damaged parts, which calls for the continuous input of energy. So living things are not really "things," they are processes; structures that capture the flow of matter and energy, like eddies in a stream.

Organisms fall into the large and heterogeneous category of "systems": entities composed of elements that interact, or are related to one another, in some definite manner.[4] A bicycle is a system, a city is another, the whole earth can be regarded as a third; by contrast, a chunk of granite is not a system, it's just an aggregate of crystals of half a dozen kinds. Cells and organisms are open dynamic systems, whose stability depends on the continuous flow of matter and energy through themselves. A hurricane is an example of such a system from the physical realm, but organisms are first and foremost chemical systems. With the advent of computers an intensive search has been launched for principles common to all systems, particularly complex ones. Such principles do exist, at least at a general level. Figure 4.1 illustrates one that is prominent in living systems: causation flows both upward from the parts to the whole and downward from the whole to its parts, but by different paths As outlined in earlier chapters, living things conduct their affairs by means of

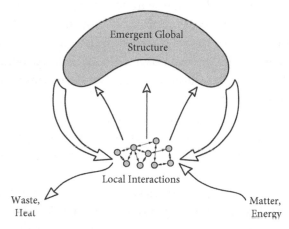

Figure 4.1 Complex dynamic systems: Causality flows both upward from the parts and downward from the whole. Modified from a source that I cannot locate.

specialized molecules whose activities are commonly regulated so as to serve the needs of the organism as a whole. Genes specify which molecules any particular cell can make (downward causation), but it is the cellular network that controls which genes are to be expressed at any moment, and to what extent (upward causation). Causality that flows both up and down is a principle of systems biology that organisms share; it lies at the heart of nonlinear responses such as excitability. But one must never forget that living systems display at least one feature that is not found in any other system: they contain a genome that specifies the structures and functions of the molecular parts, and that genome's information content is not subject to control by the cell as a whole. Organisms really are systems in a class by themselves.

There is something disarmingly old-fashioned about "organisms." The term evokes the ghost of Immanuel Kant, the German philosopher who pointed out their essential nature two centuries ago: In organisms, not only do the parts work together (as they do in a machine)but also they produce the organism and all its parts. Is such a fusty term still useful in an era when computer-savvy scientists manipulate genomes at will and look forward with mingled glee and apprehension to synthetic forms of life? Indeed it is, and for anyone concerned with what life is and how it came into existence it is indispensable. After all, organisms make up the form in which the phenomenon of life is dispensed; and it is organisms, not genes or molecules, that reproduce, act on their surroundings, and become targets of natural selection. Nothing below that level can be said to be alive (not even viruses). But the concept is actually quite slippery, and it can be hard to make out just what makes an organism. A cell of *Escherichia coli* seems clear enough, a unicellular organism, but what about a mouse? The wee timorous beastie that we are familiar with is actually a community in which bacterial associates far outnumber their singular host. A liver cell, in turn, is an organism in one sense and at the same time a subordinate member of a hierarchy. Lichens and corals are products of symbiosis, whose partners can also live independently. Evidently, living things vary in the degree to which integration of the parts prevails over conflict between them. "Organismality" is a continuum, not an all-or-none principle, that embraces both the bacterium and the lichen. What unites them is that organisms act in a purposeful, goal-directed manner, and their parts have functions. The essence of organismality lies in that shared purpose, the welfare of the integrated whole.[5] So organisms are units of adaptation as well as selection; without them the idea of "life" would cease to make sense.

Is DNA the Blueprint for a Cell?

The discovery of the structure of DNA, and of the mechanisms responsible for the expression and replication of genetic information, made an enormous impression on all who think about life. Today biologists march under a banner emblazoned with the double helix, that proclaims, "DNA makes RNA makes protein." The experimental findings were soon generalized into a far broader doctrine: all of heredity depends on DNA, which specifies all the traits and activities of living things. DNA has come to be seen as the master molecule, the essence of life itself. Francois Jacob, early in the heroic phase of molecular biology, famously wrote, "The whole plan of growth, the whole series of operations to be carried out, the order and the site of synthesis and their coordination are all written out in the genetic message."[6] Most scientists would agree with Renato Dulbecco that "life is the execution of the instructions spelled in the genes,"[7] and the public has swallowed the doctrine whole. Thirty years ago, the human genome was touted as the very blueprint of human life. And there is truth in this, important truth. Genetic information, its transcription, and its translation are indeed central to heredity and to the way life works. The question is not whether DNA is important but whether it is the sole fount of biological order, and that turns out to be a rather more subtle question that hinges on the spatial organization of cells.

Anyone who has watched a dividing cell (or merely tried to follow the plot in a biology textbook) will have been impressed by the intricate choreography. Imagine reproducing an airliner the way cells divide, and you get an inkling. Duplicating the molecules, millions of them of thousands of different kinds, is the easy part. Cells must also reproduce the pattern, the arrangement of those molecules in space—everything from ribosomes and membranes to nuclei, organelles, sites of synthesis, and genomic instructions too. All the while the cell as a whole remains intact and functional. Even bacteria, so small and featureless when inspected under a microscope, turn out to have a characteristic form and quite an elaborate spatial organization including a nucleoid and a cytoskeleton. How much of this organization and its reproduction is spelled out in the genes? Does the genome specify cellular architecture, and if so how are the instructions carried from a linear sequence on the scale of nanometers to a three-dimensional body at least a thousand-fold larger? As usual, bacteria as the simplest of cells make a good vehicle for reflection.[8]

A set of remarkable experiments from the laboratory of J. Craig Venter throws the issue into sharp relief.[9] Very briefly, Venter's team replaced the genome of one bacterial species with that of another species, and documented that cells which issue from the transplantation belong to the species that donated the DNA. Working with *Mycoplasma*, which has a particularly small genome and other virtues, they isolated the complete genome from one species (here designated D, for Donor) in the form of pure, naked DNA; transferred it into the cells of a second species (R, for Recipient); and then removed the original genome of the Recipient cells. All cells of either the D or the R species were killed off during the procedure, and only the very few survived in which the genome from D was successfully transplanted into R cells. These (about 1 in 150,000) grew into normal cells that were unambiguously of the Donor species. By the time the transplants had multiplied into a colony, about a million-fold, no evidence of residual R genes or gene products could be detected. Taking the observations at face value, they proclaimed that cell form and functions are ultimately determined by what the DNA prescribes, without any contribution from the cytoplasm or cell structure. In subsequent work the team synthesized the entire genome chemically (a major feat in itself) and transplanted it into R cells, which were then enucleated. Again, the few cells that grew out were of the D species. Most recently, the team used these procedures to create a "synthetic" cell with a genome somewhat smaller than that of any natural *Mycoplasma*: a mere 473 genes, which is presently the best approximation to the smallest gene set that can sustain cellular life. Even so, the function of a third of these genes is unknown.

Can we then safely conclude that all the instructions needed to build a cell are encoded in its genome? Not so fast, because read-out of that genome evidently requires a functional cell at all times. Successful genome transplantation requires Donor genes and gene products to replace Recipient genes and gene products. We do not know just what transpires during genome transplantation, and given the very low rate of success we may never know. Note, however, that the Donor DNA is introduced into intact cells, and the Recipient DNA is removed subsequently; how much of the cell's original organization survives all through the procedure is not clear. Some of the reasons why readout of the genome requires a working cell are obvious. DNA by itself is inert, so readout requires the cell's housekeeping functions: enclosure, energy, building blocks, ribosomes, the apparatus of gene expression, and so on. The interesting question is whether the cytoplasm, or cell structure, also makes a more direct contribution to the propagation of order. The answer, it appears, is Yes.

When cells divide and multiply most of their components are made afresh from gene products, but in some cases spatial organization is transmitted thanks to the structural continuity of the daughter with the mother cell. The single most important of these is membranes; in the genome transplantation experiments, that would be *Mycoplasma*'s plasma membrane. There have been numerous reports, some going back decades, to indicate that membranes are never made afresh; they grow by extension of a previous membrane. The molecular mechanisms that insert proteins into membranes with the correct orientation simultaneously ensure that every membrane retains the spatial orientation of its parent. To be sure, those proteins themselves are specified by genes, and the lipid matrix is produced by genetically specified enzymes. But the organization of these elements into a working membrane is carried over by "structural heredity," because the growing membrane is at all times continuous with that of its parent. It is partly inherited and partly made afresh.[10] (Let me add that the assertion that spatial organization is transmitted by structural heredity rests not on proof positive, but on the absence of any known exceptions. I hope that the program on which Venter's team has embarked will supply an experimental test of this claim.)

We have already encountered a specific example of the critical importance of membrane architecture in Chapter 3, when we spoke of energy transduction by a proton current. All the chemical reactions are effected by proteins, which are themselves specified by genes. But there are no genes for oxidative phosphorylation as such, only for its protein effectors. The physiological process is a function of the system as a whole. It depends not only on having the correct proteins but also on having them all together in a single closed vesicle and with the correct orientation. It is an "emergent" property, manifested by the collective but not discernible in any of its components.

Where does this leave us? The genome transplantation experiments confirm and reinforce the long-standing conviction that traits associated with particular genes are largely determined by those genes. Genes specify all that makes one organism different from another. But there is good reason to doubt that the spatial organization of cells, the common ground of cell biology, is specified by genes, either individually or collectively. On the contrary, it appears that much of that organization is passed from one generation to the next by mechanisms wholly or partly independent of the genes. If there is something equivalent to a cellular blueprint, it is not DNA alone. We clearly need a more flexible model, one that allows for input from multiple sources rather than solely from the genome. A closer look at how cells actually reproduce supports a more sophisticated conception, one that sees cell organization as the emergent property of a complex, dynamic, and spatially ordered system composed of gene-specified elements.

What's a Gene, Anyway?

Genes are so central to our understanding of life that it comes as something of a shock to realize how fuzzy the concept actually is. When Wilhelm Johannsen coined the term in 1911, nothing whatsoever was known about how heredity works. Johannsen envisaged an entity that is transmitted in accordance with Mendel's laws, and is responsible for a particular trait or characteristic. As knowledge grew and evolved, so did the "gene." In common speech we still use "gene" to mean a heritable unit responsible for a function, as in the gene(s) for lactose metabolism, which allow us to consume milk. Confusingly, many genes are named for variants that are dysfunctional: there is no gene for lactose intolerance, or for cystic fibrosis; these are defective variants (alleles) of genes that have a proper role in physiology. Molecular scientists have to be more specific: A gene is a stretch of DNA that codes for a biochemical product, and the genome is the set of genes associated with a particular organism. This is the sense in which the terms are used in this book.

When one looks more closely, the sharp outlines of definition begin to shimmer.[11] As a rule the first product of gene expression is a molecule of RNA, which may have a function of its own (e.g., ribosomal RNA) or may be translated into a protein; and every gene has a single product. But there are instances in which a gene can be expressed in multiple ways, and specify two or more products. Conversely, recombination sometimes generates proteins composed of the instructions from multiple genes. And what about the numerous loci in a genome that do not code for anything, but perform important functions as binding sites for regulatory proteins or for regulatory RNA: Is the stretch of DNA that makes up that site also a gene? No, because it does not encode a product, but it is clearly part of the genetic instructions.

I stated earlier that to a first approximation genes specify the chemical structures of biological molecules. This assertion really needs to be qualified. Most biological substances are products of enzymatic activity, sometimes of a series of enzymes that ultimately produce large molecules such as carbohydrates. Genes do specify the sequences of the enzyme proteins, but the final product is commonly a very indirect output of what is spelled in the genes. For complex substances such as the bacterial cell wall, the gap between the product and the genes is quite enormous.

Going on to eukaryotes, matters become even more complicated. Eukaryotic genes are commonly studded with "introns," noncoding segments that have no function of their own and must be spliced out before translation. They may be remnants of parasitic elements that have decayed and become locked in place, and sometimes make up more than 90% of the genome. Junk DNA? Maybe, but much of this apparently useless DNA appears to be transcribed and may be physiologically useful in ways that we do not yet understand.[12]

The genome is not only a key component of the living system but also it is itself a complex system whose elements cooperate and also compete; they promote and also constrain each other in ways that are very hard to unravel. On the scale of a lifetime genomes are durable, but on the evolutionary time-scale they are seen to be malleable, almost fluid.[13] Advanced processes, like cell division or migration, are outputs of multiple interactions; to single out any one of the genes involved as being the "gene for" that function is as likely to confuse as illuminate. Truly, "genomes are wonderful, crazy, complex, messy things."[14]

The Emergent Cell

"Emergence" is another of those slippery concepts that give philosophers heartburn. For present purposes it is sufficient to recognize that when two or more entities are combined into a higher one, the properties of the new entity are commonly not predictable from those of the components. Think water: do the properties of hydrogen and oxygen gas foretell wetness? This is generally true of systems, which typically have "the peculiarity that the characteristics of the whole cannot (not even in theory) be deduced from the most complete knowledge of the components."[15] Biology, a science of complex systems, supplies innumerable examples with novel properties appearing at every transition in the unfolding of life, from molecules to cells and on to organisms, societies, and ecosystems.[16]

Morphogenesis, the production of form and organization, is the grand example of emergence at the level of molecules and cells. Bacteria, like all other living things, are not shapeless blobs; they take characteristic forms (rods, cocci, filaments, and others) that are easily recognized under the microscope, and reproduced with high fidelity from one generation to the next. They display substantial spatial organization, including multilayered cell walls and internal membranes, also appendages such as flagella, and these too are reproduced time after time. Are these recurrent traits spelled out in the cell's genome? Well, yes and also no. Cells and all their parts are made of biomolecules such as proteins, nucleic acids, and polysaccharides, whose chemical structures are more or less directly specified by genes. Mutations in these genes often result in secondary changes in cell morphology and organization. But genes chiefly and directly prescribe only the chemistry of biomolecules, not the higher levels of spatial order. Between what is spelled in the genes and what one sees under the microscope sprawls a gap of three orders of magnitude, which is bridged by a hierarchy of physiological processes that are not specified by any genes. Instead, they come under the heading of self-organization: processes that generate spatial patterns on a supramolecular scale by molecular interactions obeying only local rules. Physiology is what allows cell organization to emerge from the chemical level, by the interactions among genetically specified elements within the context of a dynamic system. This is why the morphology of even the simplest cell cannot be predicted by reading its genome.[17]

Let's put some flesh on these dry bones. In the past two decades our knowledge of how bacterial cells divide has grown by leaps and bounds, and while many details remain to be worked out, the principles are clear enough. Bacterial cells do not transcribe and translate genes, and then wait for the products to fall into place. They grow, constructing the new cell on its parent such that the new is architecturally linked to the old. Figure 4.2 illustrates schematically how E. coli does it. A dividing cell starts out as a tiny cylinder about 2 micrometers long and 1 micrometer in diameter, with hemispherical caps, not unlike a miniature propane tank. This shape is dictated by the strong and rigid cell wall, which resists the osmotic pressure of the cytoplasm and keeps the cell from bursting. The wall is made up of thousands of subunits, cross-linked into a single giant macromolecule in the shape of the cell. Morphogenesis is first and foremost a matter of enlarging the cylinder, retaining its form and dimensions while maintaining its integrity at all times. This is accomplished by intercalating freshly made subunits into

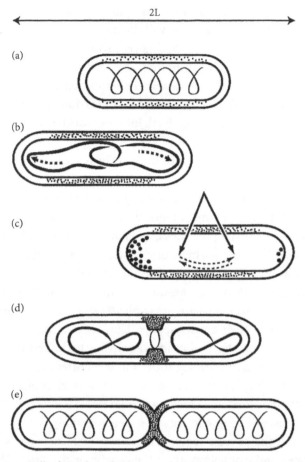

Figure 4.2 One cell of *E.coli* making two—localized and oriented processes.
(a) Actin-like cytoskeleton and dispersed synthesis of side-walls; (b) Division
of the nucleoid and active segregation of daughters to the poles; (c) Cell finds
its midpoint by oscillation of Min proteins; (d) Construction of the division
machinery at the midpoint, with wall synthesis focused there; (e) Construction
of the septum. From Harold 2005, with permission of the American Society for
Microbiology.

the existing fabric, one subunit at a time, by localized cutting and splicing.
The chief driving force for cell expansion is the osmotic pressure, but the
work is done by a battery of dedicated enzymes whose activities are localized
by the cell's cytoskeleton (the details are still rather murky). So the cylinder
grows longer, and when it has reached twice its initial length the pattern of

wall synthesis changes: elongation ceases, and synthesis comes to focus on the construction of a septum that will divide the cell in two, usually at the midpoint. Finding the middle is itself a very sophisticated operation, whose details vary from one species to another. The newly formed pole is bowed out by turgor pressure into the characteristic hemispherical cap. In the meantime the genome and plasma membrane have also doubled, and both mother and daughter cells are ready to go forth into the world. Reading this, one can hardly resist interjecting, "And how does the cell know how, when and where to do all this?" Well, in some cases the answer is known, in others not—at least not yet.[18]

Where do we find the instructions that orchestrate the performance? Many go back to the genome, others are distributed across a hierarchy of localized and directional processes as sketched in Figure 4.3. Genes specify the structures of all the participating molecules as outlined in Chapter 3 (often quite indirectly), and they are also deeply enmeshed in regulating the timing and magnitude of their production. The genes' writ extends still further, by means of effects that are not explicitly spelled out in any sequences but are implicit in their consequences. Consider ribosomes, those clever machines that zip together amino acids according to the genetic instructions. Ribosomes have a characteristic form and structure; each consists of two subunits, some fifty protein molecules and several molecules of RNA. Remarkably, ribosomes arise by the orchestrated assembly of their constituent molecules, one by one in regular order. Assembly is not spelled out in any genes, but the process requires the macromolecular parts to have the correct shape, and that is implicit in the genetic instructions. The assembly of ribosomes is one instance of self-organization, a process that generates higher-order structures by molecular association without any further input of information and sometimes even without input of energy (ribosome assembly does require energy).

Self-organization, with or without an input of energy, is a major element of cell morphogenesis whose full dimensions have yet to be clarified.[19] The apparatus of cell division arises in this manner, and so do flagella, lipid-bilayer membranes, the cytoskeleton, and many others. Many self-organized structures are dynamic, dependent for their stability on a constant input of energy. The critical feature that marks them all is that self-organization proceeds without guidance from the genome beyond that required to specify the parts, or from any other source of information. It works, but just how remains to be spelled out.

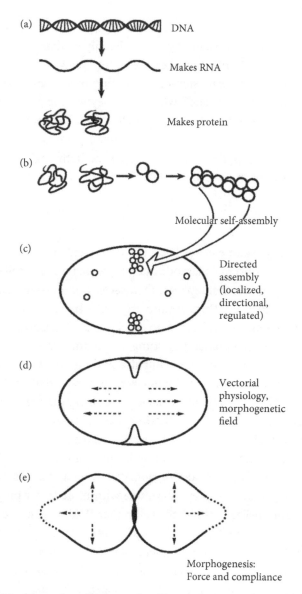

(a) DNA

Makes RNA

Makes protein

(b)

Molecular self-assembly

(c) Directed assembly (localized, directional, regulated)

(d) Vectorial physiology, morphogenetic field

(e)

Morphogenesis: Force and compliance

Figure 4.3 The hierarchy of living order. The gap between the scale of molecules and of cells is bridged by a nested series of processes, chemical ones followed by physiological ones.

From Harold 2005, with permission of the American Society for Microbiology.

So can we consider the cell as a whole the product of self-organization? Yes, in a subtle way, but now we must go considerably further down the slippery slope of complexity. As mentioned earlier, membranes are never constructed de novo, they arise by extension of preexisting membranes. That is an instance of structural inheritance, which goes beyond the remit of any genes. Another such is the localization of organelles, which is commonly specified by spatial markers carried over from the previous generation. A third factor is mechanical force, applied in the right place and in the right direction by an oriented cytoskeleton. So cell organization is partly constructed anew and partly inherited, directed by the genes that specify the parts and by the architectural continuity that links every cell generation to both its parents and its offspring.

The foregoing argument was laid out in the context of prokaryotic cells, but it applies a fortiori to eukaryotic ones. Here we find organelles, the mitochondria and plastids, that replicate themselves but do so in a manner integrated with the host cell. There are also organelles that arise by self-assembly, but whose location and orientation in cell space are transmitted by cell continuity; and others that undergo splitting. No, putting the cell in order is not simple, it's not wholly in the genes but also relies on the transmission of architecture, and there is no straightforward program that orchestrates the performance. Only the organized structure can reproduce itself, and once that structure has been destroyed it cannot be regenerated.

On the face of it there seems to be a glaring conflict between the geneticist's understanding of cell organization and the physiologist's. The former proclaims that form and organization obey the genes' dictates. The latter sees the genome as a key subroutine within the larger program of cell operations, and insists that it is the cell rather than its genome that grows, reproduces, and organizes itself. They can't both be true—or can they? In my view, these points of view are complementary, not opposites. Reproduction and evolution operate on different timescales. A growing cell relies on self-organization to transmit much of its spatial order, by mechanisms that are commonly independent of the genetic message. But genes have long-term permanence that cell architecture cannot match. So genes specify the parts, the parts instruct the whole, and on the evolutionary timescale it will be the genes that chiefly shape the cell. Having said that, there remains a long stretch between the straightforward determination of a protein's amino acid sequence by the nucleotide sequence of its cognate gene and the devious and cryptic manner in which the genome can be said to specify the whole cell.

We must not minimize the conceptual shift from a linear chain of command to a branched and braided loop of causes and effects reverberating in a self-organizing web. The only agent capable of interpreting the genome of *E. coli* as "a short rod with hemispherical caps" is the cell itself; and that is why it still takes a cell to make a cell.

Schroedinger's Riddle Revisited

Let me summarize the foregoing chapters by returning to the question, So what is life? In the context of natural science the word is used in two senses. "Life" is shorthand for all living things, as well as those that have lived in the past or that may live in future. "Life" is also the characteristic quality or attribute of organisms, natural objects that we recognize as living. Organisms are dynamic systems of prodigious complexity, composed of huge numbers of molecules structured and organized in space; societies of molecules, not mere aggregates. Cells are the basic units of life, and each cell is itself an organism. All cellular mechanisms are molecular, but it is spatial organization that brings the molecules to life. Organisms draw matter and energy from their environment, and construct themselves. They maintain their integrity even though their constituents are continuously broken down and resynthesized, they reproduce their own kind, and evolve over time by the interplay of heredity, variation, and natural selection. In all these activities DNA serves as an indispensable database, but it does not direct the show; most of what organisms do is the output of a complex, interactive system, not the product of a linear chain of command. Organisms are enormously diverse, yet all living things on earth are variations on a single molecular theme and share a common ancestry. In this sense, Schroedinger's riddle has been read: we know quite well "what life is."

This is not to say that biologists can now shut up shop: plenty remains to be discovered, even at the most basic level. That includes challenging questions about how life works, especially in higher organisms. How does an egg turn into an organism? How can we recall some fleeting experience decades later? How does self-awareness emerge from the cacophony of neural communications? We shall look at some of these matters in Part III. For the present, let me turn to the question of how living things came to be just as we find them. Even though living things are material objects bound by the laws of chemistry and physics, they are also creatures of evolutionary history and laced with a pronounced strain of contingency. They might have turned out quite differently. This is why biology will always remain an autonomous science, with its own vocabulary and principles that set biology apart from the physical sciences. While the principles of evolution are well understood, the actual history of life presents numerous problems, particularly in its early phases. Behind those looms the perennial mystery of life's origin, the black hole at the heart of biology. Until we solve that riddle we cannot rigorously exclude the possibility that life owes its very existence to forces that fall outside natural science as we presently understand it. Living things remain strange objects, not fully accounted for.

Students of life labor under a limitation that has no remedy as yet: we know only one single kind of life, the kind that is all around us and to which we belong. Is life unique, a feature of our particular planet? Given the vastness of the universe and the abundance of planets, that seems unlikely; uncommon, probably, but not unique. If indeed life on earth is but one instance of a phenomenon that is widely distributed, what features of terrestrial life should we expect to find on other planets far away? We can only speculate about that, and will do so in the Epilogue. But I feel quite confident that extraterrestrial life will also take the form of organisms: complex systems made up of numerous interacting parts that act in pursuit of a shared purpose. They will draw matter and energy into themselves, maintain their identity, reproduce their own kind, and evolve over time.

PART II
THE WEB THAT WEAVES ITSELF

5

The Darwinian Outlook

A theory is the more impressive the greater the simplicity of its premises, the more different kinds of things it relates, and the more extended its area of applicability.

—Albert Einstein[1]

Even in the restricted context of natural science, the term "Life" has multiple meanings. Thus far I have used it to designate the signature quality of molecular systems of daunting complexity that draw matter and energy into themselves, maintain their identity, reproduce their own kind, and evolve over time. But "Life" also serves as shorthand for the totality of all such systems that make up so much of our daily experience: the plants, animals (including humans), even microbes. It also covers similar objects that lived in the past and are known to us only from fossils—mammoths, dinosaurs, ammonites, and bacteria of a billion years ago. Life in this collective sense offers a perspective complementary to that focused on individual organisms, like the difference between a forest and the trees. It is true today, and has probably been true from the beginning, that organisms never occur singly nor in aggregates of a single kind. They invariably appear as members of a community or ecosystem whose citizens differ widely in size, form, and lifestyle, but whose habits and lives are intertwined. How did the dense multicolored tapestry of living forms come into existence? That is the question that will occupy us for the next three chapters.

Through Darwin's Eyes

There are ultimately only two ways to account for the existence of life. One postulates an act of creation on the part of one or more gods, so that life reflects the mind and will of its creator(s). This belief has satisfied generations of people, and millions continue to find meaning in it today. It has inspired painters, poets, philosophers, and scientists, and it lies at the root of all Western societies. Only a minority embraces the alternative view, that the living world owes nothing to divine intervention but is wholly the product of natural causes acting over billions of years. The theory of evolution, articulated by Charles Darwin in 1859 in a book titled *The Origin of Species*, shook up European society in its day, and continues to reverberate in our time.

Darwin's insights were simple, but their explanatory power is enormous, and they remain the core of the evolutionary outlook. Similarities among organisms are chiefly the result of common descent. Organisms pass their characteristics to their offspring with high but not perfect fidelity. Accidents happen, mistakes occasionally occur during reproduction, and some of these are passed on to the next generation, to be inherited in their turn. Most variations are harmful and are soon eliminated by the death of their bearer, but a few turn out to be beneficial in some manner and make their bearer more likely to survive and leave successful offspring. Natural selection, the very key to Darwin's theory, does not represent the verdict of some external agency that rewards good choices; it is simply that the more effective organisms are more likely to persist, flourish, and multiply. The result is that the characteristics of organisms are not fixed, but change slowly over time. Those living today diverged from their ancestors on a timescale of millennia. This process has allowed life to keep up with changing circumstances, generating a host of useful adaptations step by tiny step: eyes to see with, flowers to attract pollinators, devices to resist disease. Darwin did not invent what we now call "evolution": the idea that organisms change over time had been under sporadic discussion for a century. What was new was the specific mechanism to drive change. Thanks to the interplay of heredity, variation, and natural selection, life has proliferated, become ever more diverse, and colonized every habitable corner of the planet. Wherever you look, even in the depths of the ocean and in hot rocks far beneath the surface, life has made its home.

Darwin's thesis made a radical break with conventional beliefs. He was quickly successful in persuading his contemporaries that organisms have changed over time and that this history can be read in the fossil record, but natural selection remained controversial for decades. Objections rained down, not only from religious quarters but also from his fellow naturalists. Is it really credible to attribute such marvels of design as the human eye, or the staggering diversity of beetles, to the cumulative effects of small variations? The age of the earth was in dispute, and while few scientists adhered to the biblical age of 6,000 years, many doubted that the earth was old enough for Darwin's plodding mechanism to work. Besides, with the mechanism of heredity quite unknown, it was not obvious how occasional variations could be transmitted from parents to offspring. Not until well into the 20th century, following the recognition of the earth's antiquity, of chromosomes and genes, and the application of sophisticated statistics, did evolution become the standard model of biology. Over time it has worked as a leaven, corroding the traditional wisdom not only in science but also in social studies, economics, and politics.

Evolutionary theory emerged from these discussions greatly strengthened and in a "hardened" form (as the late J. S. Gould aptly put it). The "modern synthesis" was formulated in the 1940s and remains today the conventional view, often referred to as "neo-Darwinism." A brilliantly successful melding of the zoological and botanical lore of Darwin's time with the rising science of population genetics, its central postulate is that genes rule. Genes specify the structure, activities, and development of living things, and thanks to their precise replication they (more precisely, the information they encode) are potentially immortal. Evolution tells the story of the genes' adventures over time. The transformation of organismal forms and functions, which had traditionally been the focus of evolutionary thinking, were marginalized, for these are but secondary consequences of what the genes mandate. Genes undergo modification by mutation and other events, and thus come under the sway of natural selection, albeit indirectly: natural selection acts on organisms and seldom sees genes as such, but selection of organisms will favor some genes (more precisely, alleles) over others, and thus shift their frequency in the gene pool. Evolution is slow and gradual, creeping at a petty pace from one mutation to another. The underlying events occur at random, in the sense that they happen unpredictably by chance regardless of need. Acquired characteristics (such as the blacksmith's muscular arms), however useful they may be to the individual concerned, are never inherited because they cannot be registered in the genome. It is this interplay between chance mutation and natural selection for better performance that has shaped all living things, generating both their overwhelming diversity and the clever adaptations that earlier philosophers had credited to purposeful design.

This, in brief outline and crammed into a nutshell, is how scientists presently understand the genesis of the living world. The proposition is buttressed by reams of evidence, and dovetails admirably with the gene-centered conception of how cells work as sketched in Chapter 4. Nevertheless, like genocentric theories in other branches of biology, neo-Darwinism has come under attack, not because it is incorrect but rather because it is too narrow. I share this opinion (up to a point), and will explore revisions to the theory in the following section.

Enlarging the Evolutionary Envelope

The modern synthesis first crystallized three quarters of a century ago, at a time when cell biology and biochemistry were just getting started and bacteria were almost *terra incognita*. Not surprisingly, the postwar explosion of knowledge turned up many findings that bear on evolution and some that challenge the received doctrine. Consider some examples:

- Molecular sequences are not the only characteristics that are inherited. Certain cellular structures pass from one generation to the next by structural continuity, notably membranes and several eukaryotic organelles.
- There is presently intense interest in "epigenetic" modifications of the genome, which do not entail changes in DNA sequences but can nevertheless be inherited. Such changes leave the content of the genetic message unchanged but modulate its expression, often with profound consequences for such processes as embryonic development.
- In classical genetics, the location of a gene relative to other genes or to chromosomal markers was assumed to be fixed. This is not necessarily true. Examples of mobile genes include various elements related to viruses ("jumping genes"), and especially the transfer of genes from one species to another.
- We still take it for granted that genes are permanently bound to the organism in which they are found. This holds true, by and large, for animals and plants but not for microbes. Gene transfer from one species to another, even from one phylum or domain to another, is a major contributor to cell evolution that will be discussed in the following chapter.

- Acquired characteristics can sometimes be inherited after all, especially in microorganisms. Examples include certain features of cell structure, and most significantly intracellular symbionts. Symbiosis is a major factor in cell evolution, particularly in the origin of the eukaryotic cell (Chapter 6). Alterations of this kind are described as "saltatory," jumps that change many characteristics at once and greatly accelerate evolutionary change.
- Can organisms rewrite their own genome to suit the needs of the moment? Must selection always take place at the level of individual organisms, or can it also happen at other levels—genes, for instance, or social groups? A case has been made for these and other drastic revisions,[2] and while they remain on the margins of the conversation it is important to be aware that evolutionary theory is itself still evolving.

Current disputes turn not so much on the findings themselves as on the severity of the challenge they pose to our understanding of adaptation, novelty, and speciation. Recall that the modern synthesis was formulated with only animals and plants in mind, and the living world has expanded greatly since then. Unicellular organisms, prokaryotes especially but also eukaryotes, practice modes of heredity that were not anticipated and that clearly contravene both the letter and the spirit of the original synthesis. That the theory needs renovating is beyond question, but should it be replaced altogether? There are serious scholars calling for a drastic overhaul, notably James Shapiro and Denis Noble,[3] but the issues are subtle and require careful consideration.

The heart of the matter is the place of genes, both in cell physiology and in evolution.

The conventional view puts genes central in both cases: the genome directs the composition of cells and all their activities, and evolution begins with random variations of the gene complement winnowed by natural selection. Regarding physiology, Chapter 4 laid out a more nuanced position which recognizes that a cell is a system, a network of elements that communicate, interact, and constrain each other. Information and causality flow both up and down, by different pathways, from the genes to cellular structures and operations and from the working system to its genome (Figure 4.1). Genes specify which molecules can be made, but the cellular system decides which genes are to be expressed and how intensely. However in that web of interactions the sequence specifications delivered by the genes carry a particular burden of *gravitas*: cells are systems constructed from parts specified by genes. Only parts spelled out in the genome can be produced by the cell. Therefore, no photosynthetic *Escherichia coli* and no flying pigs. There is nothing else quite like living systems, and no satisfactory analog or metaphor.

This point of view can reasonably be applied to evolution too, keeping in mind the differences in scale. Physiology scrutinizes the workings and responses of individual organisms and operates on the scale of the life-cycle. Evolution is concerned with changes in populations over hundreds of generations and revolves around history, the origins of novelty, adaptations, mounting functional organization, and the boundless diversity of living things. But both rest on the same pillars: a framework of gene-specified elements that interact with and constrain all the activities of life.

We now understand that genomes are neither fixed nor isolated; they get tweaked, trimmed, pulled to pieces, shuffled, and trafficked. Genomes are subject to many more influences and insults than mutations alone, and on the millennial timescale are seen to be in a state of ceaseless flux. Sometimes what James Shapiro calls "natural genetic engineering" is part of normal cell operations (as in the switching of surface antigens in bacteria, or of mating types in yeast), but it can also bring about lasting and profound changes. For example, proteins are commonly constructed of distinct functional domains. The segments of DNA that encode these domains are not products of spot mutations but were borrowed as a block from other genes, with the aid of the virus-like entities called transposons. The shuffling of protein domains creates novelty at a bound, not at the petty pace of mutation. Prokaryotes, in particular, are now known to evolve by a combination of processes that includes both traditional mutations and the acquisition of genes from other organisms. The most consequential of the nontraditional modes of heredity is symbiosis, especially the permanent intracellular associations that create new kinds of organisms. This is how mitochondria and plastids arose, and probably that most spectacular of innovations—the eukaryotic cell itself. Changes at either the level of genes or of the cellular system can potentially serve as starting points for subsequent evolution, whose course will be governed by both natural selection and by the system as a working whole.

Having said this, let me quickly put on my other hat and stand up for the special status of the genes. Genes, and only genes, are copied precisely generation upon generation. The information they embody is potentially immortal, and in practice remains recognizable after billions of years. For all its malleability by insults of every kind, genetic information is far more durable than that built into cellular structures and operations. To what extent acquired characteristics, variations at the physiological level, can be integrated into the gene complement (what Conrad Waddington, seventy years ago, dubbed "genetic assimilation"), is not altogether clear. I am not aware of an unambiguous example from multicellular organisms. In any case the iron-clad principle appears to be that variations, whatever their origin, can only endure over the long haul of thousands of generations if they are inscribed in the genome. Living things guard that invaluable database by elaborate mechanisms to minimize errors and correct those that slip by. The genome is not merely "a tool of the cell," as Denis Noble and James Shapiro insist (quoting Barbara McClintock). To the best of my understanding, cells cannot routinely rewrite genome sequences in a directed, specific manner to cope with stress or take advantage of an opportunity (the isolation of genomes from casual tampering is why one can use molecular sequences to construct phylogenetic trees). Cells do commonly possess mechanisms to reshuffle their genome or to manipulate the frequency of mutations, and they call on those in case of dire emergency. But these are still chance events, not specified alterations, and they can only make it into the population at large via natural selection. At the end of the day, evolution, with all its marvels and oddities, depends on errors that slipped through the cracks. Organisms do not pull themselves up by their own bootstraps.

It is surely no accident that the procedures of gene replication, transmission, and expression are fundamentally universal. All living things do it with polymerases, messenger RNA, ribosomes, and self-folding protein chains, with many variations of detail but on the same principles. The more things changed over the geological eons, the more they also remained the same. Evolution does not work quite as Darwin and his successors imagined, and even the modern synthesis now appears to be but one stage in the continuing refinement of evolutionary thought. But we must never forget that as long as the narrative revolves around heredity, variation, and natural selection, we are walking in the path that Darwin first blazed.

Organization without Design

Darwin's theory was intensely controversial in the 19th century, and remains so in the 21st. Not among scientists, who all but unanimously support the principle of evolution by heredity, variation, and natural selection, but among the public at large. If the polls are to be believed, about half of respondents in the West reject evolution on principle, and in non-Western cultures the proportion is downright lopsided. Why, despite all the evidence in favor of evolution by natural causes and the lack of any evidence for divine guidance, have we failed so miserably to carry the public with us?

Humans, it seems, are programmed to seek intent, purpose, and significance in events.

This, I suppose, is why conspiracy theories flourish whenever something happens, while the truth struggles to be heard. Traditional attitudes to nature mesh smoothly with our proclivities, while evolution by chance variation and selection for reproductive advantage does not. Living things flaunt endless features that are obviously there for a purpose. Is it not plain as a pikestaff that they are products of design? In a book titled *Natural Theology*, William Paley (1743–1805), one of Darwin's sources of inspiration, laid out the argument forcefully and eloquently. Should you stumble upon a watch on the wild heath, even if you did not know what this device is good for, you would be in no doubt that it was made by a skilled craftsman for some purpose. By the same token the intricacy and perfection of living design proclaim it to be the work of a creator, and make powerful evidence for his existence.

The argument for intelligent design continues to put forward in our day,[4] and clearly resonates with many people. Setting aside the more preposterous claims of biblical fundamentalists (who get hung up on the literal reading of scripture and the age of the earth), the postulate of an intelligent and engaged deity is not inherently antiscientific; after all, two centuries ago it was the norm, even among scientists. Moreover, some of the issues raised by the more serious advocates of design are not without merit. Today, scientists (including myself) overwhelmingly reject the argument for design. But it should not be overlooked that we rely on random variation and natural selection to bring forth purposeful and effective organization, quite as though an intelligent creator had decreed it.

Our confidence in the naturalistic interpretation of life is grounded in both reason and numerous details of anatomy, physiology, molecular sequences, and fossils; it is solidly rooted. But many fundamental questions about the nature of this process remain open, and should not get lost in rancorous and politicized disputes. Do random mutations, chance associations, and competition suffice to account for the endless diversity of living things? Is anything else required to explain the large-scale direction of evolution? Do Darwinian mechanisms account for the odd history of cells, and especially for the initial appearance of functional order at the origin of life? Is it true that life is wholly the product of blind algorithmic events, devoid of direction, purpose, and larger meaning, or does it shape its own destiny? These are legitimate questions, and they will lurk in the background throughout the remainder of this book.

Box 5.1 What's Wrong with Intelligent Design?

Cards on the table: Like most biological scientists I am firmly committed to the Darwinian outlook, and reject the proposition that living things are products of "intelligent design" (ID). That dichotomy has become a shibboleth, dividing folks who think about such matters into warring camps that can barely speak to one another in a civil manner. Belligerence may boost one's self-esteem, but it does a disservice to the larger public that is not pledged to any ideology and just wonders about life and how it came to exist. ID endows the world with purpose and meaning, and therefore connects more smoothly with conventional thinking than does evolution by random variation culled by natural selection. From time to time I receive a plaintive e-mail asking just what is so wrong with intelligent design, and the senders deserve a straightforward reply.

Proponents of ID commonly present highly colored claims that evolution cannot account for this or that feature of life, particularly life's origin, and must therefore be rejected as an explanatory principle. It is true that the origin of life, and also consciousness, have stubbornly remained mysterious. But that is no reason to discard the evolutionary approach. Over the past seventy years we have witnessed an explosion of knowledge, and seen one mystery after another yield to evidence-based inquiry. The origin of life is the toughest nut of all, but in the long run I'll place my bet on science to crack it.

In the meantime we should inquire just what the ID hypothesis proposes, and what it purports to explain. The central assertion is that the undirected interplay of random variation and natural selection does not, and cannot, account for the emergence of functional design, purpose, and mind from inmate matter—at least, not unaided. These glaringly obviously qualities of life testify to the involvement of some external force that lends direction to biological history. Proponents of ID differ over precisely what they advocate, but most (like William Paley) sense behind the appearances the mind and will of some higher power, the "Designer." For most adherents that power correspond to God as conceived in the Christian tradition, but it need not. Give rein to your imagination, and play with the fantasy that a team of savants at the Intergalactic University, four billion years ago, synthesized a basic life form and seeded it onto a virgin earth as a long-term experiment in evolution; that, too, would be ID.

Can science reject such cogitations out of hand, as being untenable in principle? I don't think it can. After all, as recently as Darwin's time it was quite generally taken for granted (even by such luminaries as Galileo and Newton) that the world was divinely created. We cannot even assert that that a role for the Designer is improbable: "improbable" compared to what? No, the reason scientists all but unanimously reject intelligent design is that its rationale and postulates very often (not always) contradict fundamental principles of scientific inquiry.

Science has become spectacularly successful by subjecting all its observations and ideas to stringent critical scrutiny. (Granted, that is an ideal. Individually scientists, like everyone else, are fallible, gullible, and corruptible. It is the community, the system, that corrects its mistakes and learns from them.) We try to formulate hypotheses so as to make them testable, or "falsifiable." We are leery of any proposition that invokes entities inaccessible to observation, experiment, or reason. It is conceivable that, as a matter of fact, an ineffable and incomprehensible Designer does exist, whose mind and will played a role in shaping life as we find it. But in the absence of evidence, any evidence, this is at best a speculation of marginal relevance to science. Over the past four centuries science has generated a vast body of knowledge that is (mostly) verifiable, comprehensible, and coherent. It leaves little room for a Creator or Designer, and few scientists still find that proposition credible.

One other factor explains why the community of science is so strongly opposed to ID.

In the past, scientists have repeatedly fought to carve out space for rational inquiry free of interference by organized religion. In the ID movement, especially as it has unfolded in America, we sniff a stealthy attempt to reimpose a religious framework, particularly on education; and we know where that road leads.

What's wrong with intelligent design is nothing less than that it contradicts the entire scientific outlook, and that of biology in particular. For biologists, evolution has become the overarching principle that makes the living world comprehensible. Indeed most of us would agree with Richard Dawkins, that evolution is the *only* philosophical principle that can do so.

6

Evolution of the Cell

We have now come to rely on gene phylogenies to recreate evolution's pattern, and to see genomes as surrogates for organisms, as well as chroniclers of the process of evolution: we have reduced evolution to (phylo)genetics and genomics.

—W. F. Doolittle et al., "How Big Is the Iceberg of Which Organellar Genes in Nuclear Genomes Are but the Tip?"[1]

Is there a grander show on earth than the stately unfolding of life over the ages? Museums of natural history proudly exhibit vestiges of past life pulled from ancient rocks—trilobites, ammonites, leaves turned to coal, dinosaurs, mastodons, and the skulls of hominids long gone extinct. The performance covers half a billion years, from the first appearance of animals about 550 million years ago to the present. Most readers of this book will be broadly familiar with the plot, and accept that the force driving the narrative is the interplay of heredity, variation, and natural selection. What came before is another country, an area of darkness fitfully lit by flashes of light. Museums give little space to the three billion years from the origin of life to the Cambrian explosion, which do not lend themselves to spectacular displays that draw in the crowds. Yet, from the standpoint of life as a phenomenon of nature, this darkling time is crucial, for it is the era of cell evolution when everything of fundamental importance first took shape; and it had largely run its course long before the first animals left their traces in the mud.

The excavation of that remote past is an ongoing effort that began well within living memory. When I was a graduate student virtually nothing was known about how cells came to be, except that fossil bacteria nearly 2 billion years old did testify to their antiquity. The first major breakthrough came in 1967, with the recognition that mitochondria and plastids (the seats, respectively, of respiration and photosynthesis in all eukaryotic cells) are not native to their location but descendants of free-living bacteria that had taken up residence in the cytoplasm at an early stage of evolution. The idea had been around for decades, but few took it seriously until Lynn Margulis (then Sagan) marshaled the evidence and promoted the cause. A decade later the subject cracked open, thanks to revolutionary developments in the stodgy practice of classification.

The fossil record is much richer today and remains indispensable, but the richest fount of information and ideas about cell evolution, by far, is the genomics of contemporary organisms; and while both raise as many questions as they answer, they have supplied a broad framework for reflection. My object in this chapter is to celebrate what we now think we know, to point out the many matters that remain uncertain, and to underscore one of the central issues in cell evolution: the genesis of eukaryotic cells.[2]

The Tree of All Life

Naturalists from Aristotle to the present have sought to bring order to the extravagant diversity of living things by sorting them into groups that share common features. By Darwin's time it was already common practice to assign organisms to bins nested within larger bins on the basis of similarities of form and function, as we continue to do today. Lions are placed in one genus, tigers in another, but both are felines. Dogs are obviously quite different, but all three go into a big bin that houses all mammals.

Darwin gave the practical art of classification a higher purpose. One of his most profound insights was that all organisms are related as members of a huge extended family, united by descent from a common ancestor; every species is a twig on a branch of a collective entity, the tree of all life. The object of taxonomy should therefore be to discover the natural classification, that which corresponds to the lines of descent. In time it should be possible to draw up a tree on which every living thing has its proper address. Partial trees that cover the animal and plant kingdoms could be constructed quite successfully, using forms and functions as criteria of relatedness; it proved much harder to envisage the tree of life as a whole, uniting organisms that apparently have nothing in common. How would one assess the relationship between trout and truffles, or petunias and pill bugs? Where do bacteria fit into the grand scheme, and what should we make of the discovery that all cells fall into two broad classes called prokaryotes and eukaryotes? These conundrums and others were entirely intractable until our own day, but have begun to yield thanks to the introduction of a new set of criteria to measure relationships: molecular sequences of large polymeric molecules, either the sequence of amino acids in proteins or that of nucleotides in RNA and DNA.

Like the organisms in which they are found, molecular sequences evolve and diverge over time. The order (sequence) of amino acids in the hemoglobins of lions and tigers is nearly identical, that of dogs is distinctly different, roundworm hemoglobin even more so. One can express the difference in numbers and state quantitatively how closely tigers and dogs are related. One can even use this distance to estimate how long ago these two mammals diverged from their common ancestor. For example, it has been about 6 million years since the last common ancestor of humans and chimpanzees roamed the forests of East Africa. Hemoglobin is a protein of mammals, seldom found in plants; indeed, no single protein occurs in all organisms. But all cellular beings contain ribosomes, the tiny machines that zip together amino acids to make proteins, and all ribosomes contain several particular species of RNA. Nucleotide sequences, like those of proteins, change over time but more slowly; they make a chronometer to track the ages. By mapping nucleotide sequences of ribosomal RNA from many organisms, the late Carl Woese (1928–2012) created the first universal tree of life in the 1980s, and in the process turned our conception of life upside down.

Take a close look at Figure 6.1, for the universal tree differs quite drastically from the way organisms were classified prior to the molecular revolution.[3] Like the modern synthesis, antebellum taxonomy revolved around "us," the higher organisms (Figure 6.1A). Living things were divided into five kingdoms, three of them multicellular: animals, plants, and fungi. A fourth kingdom housed the unicellular protists, all of them eukaryotic, such as algae and protozoa; and basal to the whole lot were the prokaryotes, kingdom Monera. RNA sequences, which lend themselves to more objective estimation of evolutionary distances, lay out a very different order (Figure 6.1B), in which the bulk of life's diversity and most of its evolutionary history revolve around microbes. By contrast, the household organisms—the plants

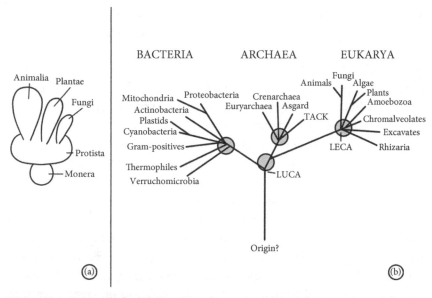

Figure 6.1 Trees of life. (A) Five Kingdoms: A traditional conception of the living world that highlights eukaryotic organisms. After Margulis and Schwartz 1998. (B) Three Domains: Contemporary classification based largely on the sequences of ribosomal RNA. Not drawn to scale. LUCA, Last Universal Common Ancestor; LECA, Last Eukaryotic Common Ancestor. Stippled areas poorly resolved. Adapted from Pace 2009.

and animal—make up mere twigs at the ends of two microbial branches. To everyone's surprise, Woese discovered almost immediately that the conventional prokaryotic kingdom Monera conflated two distinct kinds of organisms that look much alike and have similar habits but are only distantly related. The universal tree therefore divides the living world into three great stems, two prokaryotic and one eukaryotic, now designated "domains": Bacteria, Archaea and Eukarya (or Eucarya). All three share common descent from a hypothetical entity, the Last Universal Common Ancestor (LUCA), about which much more will be said in the next section. Note the absence of viruses, which are not cellular, lack ribosomes, and therefore find no home on the tree, but are somehow part of the biological universe.[4]

The universal tree transformed biology. No wonder that many classical biologists looked askance at a scheme that relegated to the margins all the creatures that have for centuries made up the core of biological science!

By now most of the technical issues have been resolved, and the virtues of molecular sequences as practical taxonomic tools are generally recognized. Zoologists and botanists were reassured to find that, for the most part, relationships inferred from RNA sequences are congruent with those drawn from classical criteria; and the universal tree has become a staple of textbooks. But the "Big Tree" is much more than a convenient instrument for sorting organisms, or even than a way of making sense of life's profusion. It is a bold hypothesis that encapsulates the entire history of life on earth, and its formulation cracked open the door into what Woese once called "the murky realm of cell evolution."

Before we go there, let me be clear about what the universal tree is, and what it is not. First, considerable uncertainty lingers over the details. The major Bacterial, Archaeal, and Eukaryal phyla correspond quite well to recognized groupings and can be taken to be real. Taxonomists are far less certain about the order in which the phyla emerged: The amount of information that can be extracted from a sequence is limited, and different algorithms yield various successions. It seems that most of these divisions emerged in great bursts over relatively short periods of time. Figure 6.1B takes the liberty of omitting the branching order altogether, so as to let the central feature stand out: life comes in three flavors, whose identities have been steadily reinforced over the past forty years by both genomics and biochemistry.

Second, the tree does not directly display the evolution of organisms, or of cells. It tracks the evolution of a single class of molecules, the RNA component of the small ribosomal subunit (more precisely, it tracks the genes that encode those RNAs). Until recently it was taken for granted that genes are permanent features of the organisms in which they reside, but no longer. We now recognize that genes undergo occasional transfer to another species, even to a totally unrelated organism! In any particular cycle of reproduction, the standard "vertical" transmission of genes from parents to offspring is the norm, but on the evolutionary timescale "lateral" (or "horizontal") gene trafficking is sufficiently common that few (if any) genes serve as indelible markers of a cell's lineage. On the millennial timescale genomes turn out to be fluid constellations, engaged in continuous swapping with a common pool of genes, the "pangenome." Gene trafficking is rampant among prokaryotes, less common among protists, and rare among multicellular organisms. The genes for ribosomal RNA are part of a small set that seldom undergo transfer (others include genes involved in the expression of genetic information, also those for ion-translocating ATPases). We take these as representative of a fairly stable genetic core that makes a serviceable proxy for cellular lineages. But a more realistic depiction would replace the tree with a dynamic network, from which one can draw a "statistical" tree of life.[5]

Finally, the premise that variations in a single class of molecules mirror all the diversity of cells and organisms is surely simplistic. Granted that ribosomal RNAs are members of a genetic core that is (almost) always transmitted from parents to offspring, that core is quite small—just a few percent of the total gene complement. Most of the genes that specify organismal features have undergone trafficking at some stage in their history, either individually or en bloc. The tree as sketched in Figure 6.1B notes two major episodes of that kind, but makes little of them: the acquisition of bacterial symbionts that gave rise to mitochondria and plastids. It follows that the universal tree sorts organisms into pigeonholes that are real and significant, but permits only limited forecasts about their traits and characteristics; and it does little justice to the rise of complexity and functional organization over time.

The universal tree of life remains a work in progress, as much aspiration as achievement. For sure, the evolutionary history of cells and organisms is too complex to be squeezed into a unitary tree based on a single class of molecules. But in coming to grips with that long and obscure past, the tree of ribosomal RNA marks a milestone, a rock on which to rest and look ahead. Battered by time and controversy, the Big Tree stands tall as a cornerstone of modern biology.

First Came the Prokaryotes

The early history of life unfolds on a timescale that boggles the imagination. Paleontologists toss around billions of years with insouciance that would do credit to a Treasury official! A billion years, one thousand million—how big a number is that? Well, suppose you are an insomniac counting sheep to fall asleep, one second per sheep; it would take you more than thirty-one years to rack up a billion. The sheer remoteness of the events discussed in this chapter sounds a warning to be a little skeptical of all assertions and inferences pertaining to a time when the world was very different from the one we know.[6]

Few rocks more than 2 billion years old have survived unaltered, and so the fossil record of earliest times is especially sparse and fragmentary; but it is clear enough in one important respect. Body fossils that look exactly like contemporary bacteria supply good evidence that prokaryotes lived as early as 3.5 billion years ago, perhaps more, and chemical markers point to the presence of both Bacteria and Archaea. What is lacking is convincing traces of eukaryotic cells or organisms. Now, the absence of such evidence does not prove that eukaryotes did not exist prior to two billion years ago. They may just have been absent from the few places we can sample, or structurally unsuited to fossilization, and therefore many serious scientists dismiss the lack of eukaryotic fossils as uninformative. I take the contrary position: One basic premise for making sense of cell evolution is the proposition that for the first 2 billion years all the life that lived was of prokaryotic grade, and eukaryotic cells did not appear until halfway through life's history. (Needless to say, this premise is not universally shared.)

The second fundamental premise is that all living things are of one kind and share a common ancestry. That ancestry is represented on the universal tree (Figure 6.1B) by the first node, the one that divides Bacteria from the Archaea, and is designated LUCA. This seminal bifurcation, the earliest event in cell evolution for which we have some evidence, probably occurred more than 3 billion years ago.

What can one say of LUCA, a hypothetical creature that no one has seen and that will have lived so deep in the past? Not very much and that most tentatively, but extrapolation from the genomes and physiology of contemporary organisms supplies meaningful clues. The presumption is that ubiquitous features (and the genes that encode them), those shared by all or most living things, are also the most ancient and likely to have been part of LUCA's endowment. These include proteins composed of the canonical set of twenty amino acids, RNA and DNA, and the genetic code. LUCA will have produced proteins on ribosomes, and specified their composition with the aid of genes. She was bounded by a lipid membrane bridged by transport proteins, and some sort of energy coupling by ion currents was in place. Potassium was already the chief intracellular cation, while sodium was expelled. Metabolism revolved around ATP and ADP, pyridine nucleotides, and other universal coenzymes. LUCA may have been relatively simple by the standards of contemporary cells, but she appears to have been a cell of sorts, equipped with as many as 500 genes. There is reason to believe that she was structurally quite elaborate, with internal membranes that were jettisoned later by evolutionary streamlining. But LUCA was probably not an organism in the modern sense. Carl Woese, who did more than anyone to clarify the concept of a universal ancestor, argued that LUCA was not even a discrete organism but a population of noncellular entities that readily swapped genes, evolved communally, and lacked the sharp lineages that define contemporary organisms.[7]

How did LUCA make a living? We really have little idea how LUCA drew energy and matter from its environment, but there are not many options. The atmosphere 3 billion years ago was quite devoid of oxygen, which precludes the respiratory metabolism that sustains most of contemporary life. Oxidation based on sulfate or nitrate is one possibility, fermentation of amino acids or other organic substances that accumulated by chemical processes on the early earth is another; some argue in favor of an early mode of photosynthesis. The most plausible idea is one first mooted a century ago: The primary energy source was geochemical, such as the reduction of CO_2 by hydrogen gas to generate methane or acetic acid. There are certain organisms, both Archaea and Bacteria, that live in this manner today, and there is persuasive evidence from bacterial genomes that the pertinent chemical reactions are among the most ancient ones to survive to the present. LUCA may have been an anaerobic autotroph that could live off gaseous hydrogen, ammonia, and carbon dioxide. The trouble is that contemporary methanogenesis and acetogenesis are both highly sophisticated processes that require elaborate biochemical machinery, and they make for a hardscrabble existence on a narrow margin. They look evolved rather than primitive. The question of the source of energy at the inception of cellular life is hotly debated, and I shall leave it there with a big question mark.[8]

LUCA does not represent the origin of life. If the foregoing arguments are correct, LUCA must have been the product of prolonged evolution, to which we shall turn in Chapter 7. But LUCA does, by definition, represent the last common ancestor from which all extant cells and organisms are descended, and a critical stage in cell history. In Woese's view, by that day evolution had reached the point when the cost of relentless gene trading began to outweigh its benefits, and discrete organisms with structural and genetic integrity performed better than the (hypothetical) free-trade zone. With the passage over the "Darwinian threshold" begins the mode of evolution that has ruled life ever since. Questions abound: it is not at all clear how we ended up with two patterns of prokaryotic organization, rather than three or more or different ones. But there are good reasons to believe that the bifurcation into Bacteria and Archaea set the stage for all of subsequent evolution.

The crystallization of the two great stems entailed numerous changes to cellular operations, and presumably took place over many thousands of generations. Bacteria and Archaea represent variations on the universal themes of cell biology (such as DNA, RNA, proteins, the genetic code, lipid membranes, ion currents). But they also display conspicuous differences that reach far beyond the ribosomal RNA sequences by which these two domains were first recognized: cell walls, flagella, ribosomes, mechanisms of gene expression and DNA synthesis, even major metabolic patterns are consistently different. Gene trafficking, which is a major force in the evolution of prokaryotes, has gone some way to blur the features that define the two prokaryotic domains. Still, Bacteria and Archaea remain different kinds of organisms, even though they look so much alike that no one suspected the cleavage prior to the advent of molecular methods.

Among the major themes of cell evolution is the discovery of all the chemical reactions that support life today, including the eukaryotes. Bacteria and Archaea differ significantly with respect to energy production: photosynthesis is found only in the Bacterial stem, methanogenesis only in the Archaeal one, suggesting that both these pathways evolved after LUCA's day (but still very early in the history of life). Other differences that must have arisen early include the chemical composition of cellular membranes, and the details of gene replication and expression. The latter display a notable feature to which we shall return in the next section: Archaeal mechanisms are strikingly similar to those of eukaryotes, but less complex. Figure 6.1B accommodates this finding by showing Archaea and Eukarya as sister taxa that share a common ancestry to the exclusion of the Bacteria, but split very early on.

About 2.3 billion years ago the atmosphere underwent a drastic change: having been mildly reducing for eons, it turned oxidizing thanks to the accumulation of free oxygen.[9] What looks like an event on the geological timescale was in fact a protracted process, whose swings and oscillations consumed millions of years; not for another billion years did the oxygen level begin to approach today's 21%. Where did all this oxygen come from? All the evidence points to microbial metabolism, specifically the advent of a Bacterial phylum called cyanobacteria, who invented a mode of photosynthesis that utilizes water as the reductant and generates oxygen as a byproduct. Now, for strictly anaerobic organisms oxygen is toxic. Some presumably went extinct, others retreated to anoxic niches where they still flourish today. More proactive organisms found ways to take advantage of the new oxidant, generating energy for their own metabolic needs. The processes called respiration and oxidative phosphorylation, sketched in Figure 3.2, sustain not only many bacteria but also most eukaryotes. The microbial world differentiated into a multitude of new species, and began to resemble that which we know. As for eukaryotes, they owe their spectacular proliferation entirely to the new oxidant. Cyanobacteria can legitimately claim to have changed the course of biological history.

Finally, where do viruses fit into the narrative? We don't know, and this remains a glaring gap in our understanding of evolution. Viruses are obligatory parasites; they are an inescapable part of our world but are not cellular, lack ribosomes, and therefore have no place on the universal tree of life. I am impressed by the argument that viruses go clear back to LUCA's day and beyond, and have co-evolved with cellular life ever since genesis. Even in that inconceivably distant past, it seems that little bugs had smaller bugs on their backs to bite 'em. Viruses persisted, proliferated, diversified, and spawned descendants that still plague us today.[10]

Emergence of the Eukaryotic Cell

About 1.7 billion years ago novel kinds of organisms shows up in the fossil record. They are relatively large, a thousand-fold larger than the bacteria that make up the earlier flora, and commonly embellished with projections or appendages. What are these objects? No one can be altogether certain, but most paleontologists take them to be the remains of unicellular eukaryotes, possibly resistant cysts such as certain algae still produce today. If that surmise is correct, these enigmatic fossils mark the appearance of the most profound cleavage across the biological world, the gulf between the prokaryotic grade of organization and the eukaryotic.

Body fossils are not the only traces to survive from the remote past. Sophisticated chemical instruments allow scientists to detect minute amounts of substances indicative of particular groups, including steranes (products of degradation of sterols), that are considered markers for eukaryotes. They suggest that eukaryotes first appeared 2 billion years ago, give and take a goodly margin. Such dates are congruent with estimates drawn from gene sequences, with the help of algorithms that convert evolutionary distances into time elapsed. All the dates are open to doubt, but they do tell us that eukaryotes, ancient as they are, are not nearly as ancient as Bacteria and Archaea. Fossils that resemble contemporary organisms, such as algae and amoebas, show up more recently, 1.0 to 1.2 billion years ago, suggesting that the invention of the eukaryotic cell as we know it was a very protracted process that spanned hundreds of million years.[11]

How and why did this radically different pattern of organization emerge, after a billion years or more of relative stasis? Let us note once more that eukaryotic cells share molecular principles and structures with prokaryotes, but differ conspicuously in size, complexity, and many specifics of molecular operations (Figure 2.1). The Bacteria and Archaea explored the biochemical realm of energy production and metabolism. The rise of the eukaryotes marks a new phase of evolution, the exploration of structural and functional complexity, that eventually produced the world of multicellular organisms, including ourselves. Astonishingly, all living eukaryotes from protists to whales and redwoods appear to be members of a single clade, descendants of a unique common ancestor. The origin of eukaryotic cells remains one of the grand issues in evolution, second only to the origin of life itself, that is presently in a state of vigorous fermentation.

As mentioned earlier, the first significant insight into cell history was the recognition that eukaryotic cells are "chimeras," products of a merger of two or more different cells. This outrageous proposition, full of implications for both physiology and evolution, provoked strenuous opposition at first but gained ground following the discovery that both mitochondria and plastids (the powerhouses of the eukaryotic cell, sites of respiration and photosynthesis, respectively) contain DNA, ribosomes, and the machinery for protein synthesis. The advent of molecular sequencing clinched the case, securely anchoring both mitochondria and plastids among the Bacteria. Today, hardly anyone questions that eukaryotic cells are products of some sort of "endosymbiosis," in which Bacteria of the phylum alpha-proteobacteria took up residence in the cytoplasm of a "host" cell and evolved into mitochondria. Algae and green plants derive from a separate, later uptake of a cyanobacterium by an early eukaryotic protist (Figure 6.2). These two mergers demonstrate the fundamental significance of evolution by association, and supply historical landmarks that frame the course of events.

What manner of beast could have served as the host for the first bacterial symbionts?

One is sure to lose friends by taking a stand on this one! For many years the conventional wisdom held that the host would have been a proto-eukaryotic cell, endowed with such hallmarks as a discrete nucleus, a cytoskeleton, internal membrane-bound compartments, and the capacity to take up particulate food. Mitochondria would have been late additions to a eukaryotic cell

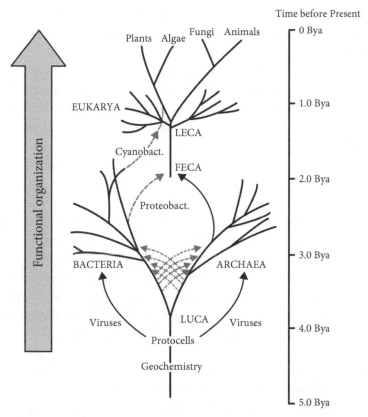

Figure 6.2 A condensed history of life. The emphasis here is on the rise of complexity and functional organization over time. Extensive interweaving of lineages by lateral gene transfer omitted for the sake of clarity. The diagram places events in chronological order, but fails to do justice to the enormous increase in complexity that the advent of eukaryotic cells entailed. Modified from Harold 2014.

whose assembly was otherwise largely complete. Indeed, there are eukary-otic protists that lack mitochondria (and plastids), and these were taken as relics of that premitochondrial phase of cell evolution. This attractive hypo-thesis had to be abandoned when molecular technology revealed mitochon-drial genes, and even structural remnants of mitochondria, in all protists that have been examined (naturally, there is one exception). The current understanding holds that all eukaryotic organisms either have mitochon-dria, or descend from ancestors that did but lost them. It is conceivable that

proto-eukaryotic cells ancestrally lacking mitochondria did exist in the remote past, but the burden of proof now rests with the proponents of that
hypothesis.

A radically different line of reasoning that has gained currency over the
past decade starts from the universal tree, but ends up in a very different
place. From the beginning, Woese and his colleagues noted the specific affinity between eukaryotic genes for ribosomal RNA and those of Archaea.
Figure 6.1 shows these as sister taxa that diverged very early in the history
of life. As genomic information accumulated it became clear that the affinity
runs both deep and wide. Broadly speaking, eukaryotic genes fall into three
classes: those that specify proteins involved in the processing of genetic information tend to resemble the corresponding genes of Archaea; genes that
specify metabolic and housekeeping activities commonly group with the
Bacteria; and many "signature proteins" are unique to eukaryotes. The most
straightforward interpretation of the observations holds that the first eukaryotic common ancestor (FECA in the trade) arose by the merger or fusion of a Bacterium with an Archaeon. Woese himself passionately rejected
this notion, and his insistence on three primary and independent stems still
has adherents. My own inclinations lean toward the merger hypothesis as
sketched in Figure 6.2, even though this formulation is by no means free of
difficulties. In one popular version of the hypothesis the Archaeon contributed much of the cellular machinery including DNA, gene expression, and
ribosomes; the Bacterium was the precursor of the mitochondrion, which
came early and played a large role in shaping the eukaryotic cell. Intimate,
intracellular associations between prokaryotes are not unheard of, but they
are very uncommon; this has been taken as evidence that the origin of eukaryotic cells depended on a very rare event, possibly a unique one. Note
that a merger wreaks havoc on the tree of life, with its three independent
stems: Instead of three there would be only two primary domains, Archaea
and Bacteria, and one secondary domain, Eukarya.[12]

A torrent of molecular evidence has been brought forth in support of
the fusion hypothesis. Some comes from re-evaluation of ribosomal RNA
sequences through the prism of novel algorithms, which show the Eukarya
branching from within the Archaea rather than as a separate stem. More
comes from the discovery of a zoo of previously unknown Archaea, the
Asgard superphylum, that live in sediments beneath the ocean floor. These
organisms could not be grown in culture, were not even seen, but their
genomes could be reconstructed by joining fragments of DNA extracted

from the mud. Some have genes clearly related to those of eukaryotes, in particular genes for proteins of the cytoskeleton and for membrane manipulation.[13] Isolation of the first Asgard archaean by a technical tour de force has just been reported.[14] The organisms are tiny cocci, half a micrometer in diameter, with conspicuous long protrusions. They live by the anaerobic fermentation of amino acids to hydrogen gas and formate, in partnership with other microorganisms that consume those products ("syntrophy").

If indeed this is what our most remote ancestors looked like we have come a long way.

The FECA may thus have been an intimate consortium between two or more grossly different prokaryotes that flourished about 2 billion years ago. She stands at the beginning of a long evolutionary process that spanned at least 500 million years and culminated in a second hypothetical entity called LECA, the last common ancestor of all contemporary eukaryotes, tentatively dated to 1.2 to 1.4 billion years ago. This immense span of time saw the gradual emergence of eukaryotic cells as we know them today, with a host of structures and processes not seen in prokaryotes but universal among the eukaryotes. Genomics suggests that LECA was already endowed with mitochondria (but not plastids), a cytoskeleton with motor proteins, internal membrane-bound compartments, a nucleus with chromosomes, and the capacity for mitosis, phagocytosis, and sex. Conversion of the Bacterial partner into an organelle (its enslavement, really) also entailed radical restructuring. Much of the Bacterium's genome was transferred to the nucleus, more was lost; and the invention of an elaborate system to import proteins brought the nascent organelle under nuclear control. Another transport system allowed the consortium to siphon off the ATP generated by respiration of the nascent mitochondrion.[15]

The genesis of the eukaryotic mode of organization was neither quick nor simple. We do not know how the assimilation began, but it seems very likely that in the long run the selective advantage that came to drive it was the gain in energy supply.[16] Remember that oxygen was even then beginning to accumulate in the atmosphere, and eukaryotes are fundamentally aerobes by nature. Much remains to be learned, and much more may remain forever beyond our grasp, but make no mistake about this: The genesis of the eukaryotic cell is the foundation for the later rise of the "higher" organisms, including ourselves. If there is any deeper meaning to the journey from the bacterium to the philosopher, beyond the endless chain of "breeding and weeding" (Steven Weinberg), it begins with the advent of the eukaryotic cell.

LECA, the (hypothetical) common ancestor of all extant eukaryotes from protists to whales, stands at the end of one chapter and the beginning of the next. That such a creature once existed is inferred from the extensive similarities of all eukaryotic molecules (and genes), which make it virtually certain that all eukaryotes share a common ancestor. Even mitochondria appear to be the progeny of one single symbiotic event. Once the eukaryotic mode of organization had come into existence, it radiated into a universe of unicellular protists that now number more than 200,000 species. Many of these live by photosynthesis thanks to an episode of symbiosis early in their diversification (dates, as usual, are uncertain, but 1.5 billion years ago cannot be far off), when members of a particular group of cyanobacteria took up residence in an unidentified protist and progressively turned into plastids, the organs of photosynthesis. As in the case of mitochondria, conversion of a free-living Bacterium into an organelle entailed extensive changes in both guests and host. The new consortium again proliferated and diversified into the algae and green plants, all of which share a common ancestry—with one exception. An obscure unicellular alga, *Paulinella chromatophora*, contains plastids still in the making that clearly derive from a separate, much more recent episode of endosymbiosis. Not only plants but also the animals and fungi can all be traced to evolutionary precursors among the protists, but that story is better deferred to a later chapter.

The Winding Road to Complexity

Let us now step back from the continuing search for a rational narrative of cell evolution, to contemplate the process as a whole (Figure 6.2). It is widely agreed that prokaryotes (and viruses) came first, followed by a long period of proliferation and diversification. The Eukarya arose much later, and they alone gave rise to all the large, multicellular ("higher") organisms. The advent of the eukaryotic cell marks a shift in the mode of evolution, a conspicuous and fateful episode that demands some kind of explanation.

Given another roll of evolution's dice, might prokaryotes have evolved into higher creatures, or is there something special about eukaryotic organization that was required to open the door? Can standard evolutionary mechanisms account for the advent of eukaryotic cells, or must we invoke extraordinary causes (as in rare, unique, miraculous, or supernatural) to make that intelligible? There is considerable dispute about these matters and little consensus, and therefore this section (even more than previous ones) reflects my personal interpretation of the facts.

There is broad agreement that the earliest event in cell history for which we have evidence is the divergence of Archaea from Bacteria, 3 billion years ago and more. But we have no clear sense of why this happened, or what selective forces drove the subsequent differentiation of the two domains. The initial bifurcation may have had to do with adaptation to energy stress, or to extreme temperatures, or to something else. What is noteworthy is that, despite much lateral gene transfer over the ages, the two domains have remained distinct; and that both Bacteria and Archaea remained small and structurally undifferentiated.

After fifteen hundred million years of relative stasis something dramatic happened, the advent of a novel mode of cellular organization. According to what is presently the prevailing view, it began with the merger of an Archaeon with a Bacterium. The Archaeon was the precursor of the genetic core, and also contributed genes that foreshadowed eukaryotic characteristics such as the cytoskeleton. The Bacterium was the precursor of the mitochondrion and contributed metabolic genes. Other eukaryotic features evolved in that lineage, including the distinctive flagellum and the elaborate system of internal compartments. Figure 6.2 underscores that the proposed merger of prokaryotes was but the first step in a protracted and convoluted process that spanned more than half a billion years.

Where prokaryotes explored metabolism and discovered all the known ways to make a living, eukaryotes tried out the possibilities of larger size and structural complexity. A few prokaryotes also ventured in that direction, but my sense is that they "always fell short" (Nick Lane). There does seem to be something about the eukaryotic pattern that renders the rise of functional complexity feasible, but there is no agreement on what that may be. I have been impressed by the argument put forward by Nick Lane and William Martin,[17] who trace the essential difference between prokaryotes and eukaryotes to bioenergetics. The prokaryotic cell is a single chamber in which energy is supplied by the common plasma membrane (and its extensions) to all the nuclei contained in the pod. Eukaryotic cells feature a decentralized energy supply derived from multiple separate power stations (the mitochondria and plastids), all for the benefit of a single dominant genome. In consequence, eukaryotes can draw on a far more abundant energy supply for every gene than prokaryotes can. The surplus power, which is the gift of "enslaved" symbionts, underpins the rise of functional organization.

During that almost unimaginable span of eukaryotic evolution, different elements contributed to the outcome: point mutations, no doubt, but also symbiosis and the loss and gain of genes in the course of organelle formation as well as by acquisition from the environment; and at least one more symbiosis that gave rise to plastids. It has been argued that the initial merger of an Archaeon and a Bacterium was an extremely rare event, possibly even a unique one, and that is why eukaryotes apparently only emerged once and after a long delay. I doubt this, and would rather believe that mergers happened several times but only one survived, perhaps because it proved spectacularly successful; or that eukaryotes had to await the appearance of a permissive environment, such as the accumulation of oxygen in the atmosphere. Still, the apparent fact that all eukaryotes are members of a single clade is not fully accounted for, and troubles the imagination.

The processes that contributed to the genesis of the eukaryotic order all appear to be quite consistent with the interplay of heredity, variation, and natural selection. If design, purpose, or a supernatural finger on the scale lurk behind the appearances, they are well hidden. I am particularly impressed by the discovery that plastids originated at least twice, once early in the line that produced the green algae and plants, and then again in the unremarkable protist *Paulinella* a mere 60 million years ago. It is hard to rationalize such redundancy on any basis other than sheer happenstance, followed by selection. This conclusion is reinforced by the failure of subsequent events to hew to any clearly defined direction. There is good reason to believe that LECA already featured all the typical hallmarks including mitochondria, nuclei, compartments, 9 + 2 flagella, phagocytosis, and sex. Once the eukaryotic order had come into existence it proliferated and diversified into multiple modes of complexity, now displayed by the variety of animals, plants, and fungi. Even single-celled protists often became amazingly intricate, as the ciliates illustrate. If we could run the tape of life over again (in Stephen Jay Gould's famous simile), it is hard to believe that the pattern of life as we know it would emerge again. The eukaryotic cell itself, and all of its progeny, appear to be products of contingent and unrepeatable events.

Well then, what quality or qualities explain the success of eukaryotic organization? Is there a "utility function" (Richard Dawkins) that natural selection has favored? Surely, this is a crucial question for anyone who seeks to understand evolution, and it is awkward that it lacks a clear and simple answer (remember that we also do not know how and why Archaea diverged from Bacteria). Eukaryotes on average do have more genes than prokaryotes do but not enormously more, and in any event there is no straightforward correlation between gene count and any other measure of evolutionary grade. Some have argued that loose and flexible control of gene expression is the magic ingredient, or introns and the shuffling of protein domains. The proposal that captive endosymbionts made all things possible by supplying the nascent eukaryotic cells with abundant energy rings true to me. There seems to be merit to all these ideas, but none is sufficient by itself to explain what it is that made eukaryotes so different, and set them on their own course.

What stands out in my mind is that elusive quality called "complexity"—of genome structure, architecture, regulation, and much else. Complexity is hard to define but easy to recognize; like pornography, we know it when we see it. Complexity is a function of the number of parts and the ways in which they interact. There is no good way to measure biological complexity, but that does not mean that it is not real or significant. Many parts, interacting in multiple ways, provide more opportunities to work with self-organization and create higher levels of order and function. Natural selection will not value (or even see) complexity as such, but it sometimes favors its consequences. Two terms that point in the right direction are "autonomy," a degree of independence from the environment; and "agency," the capacity to act on one's own behalf (Chapter 8). Such broad terms lack the specificity many people crave, but they give one a sense of the direction in which eukaryotic cells and organisms have advanced.

Complexity, autonomy, and agency come in different grades and take multiple forms. In the absence of a quantitative metric, one cannot say whether dinosaurs were more complex or autonomous than the cycad trees among which the beasts roamed. Nor would I assert that complex is necessarily superior to simple: prokaryotes are doing just fine, thank you, and many parasites have discovered that it is a gift to be simple. But complexity seems to be the way of the eukaryotes, and no one can doubt that it proved to be a spectacularly successful way to be in the world.

7

The Perennial Riddle of Life's Origin

It is increasingly clear that if we wish to uncover the fundamental nature of biological phenomena, such as function and complexity, we
will need to discover how these biological phenomena could have
emerged naturally from chemical systems. We need to understand
the physicochemical process by which chemistry became biology.

—Addy Pross, "The Evolutionary Origin of Biological Functions
and Complexity"[1]

For anyone who seeks to understand life as a phenomenon of nature, the
vital question is the origin of life. The issue invariably comes up in any conversation about life or evolution, and for good reason. The few laypersons
who take an interest in these matters appreciate quite well that living things
are drastically different from rocks, air, or water, and they look to science
to explain how that came about. But the subject of origins makes biologists
uncomfortable, for it calls into question one of our most basic insights. In antiquity and the middle ages it was accepted without dissent that rotting meat
generates maggots, and piles of rags breed mice. With the rise of science such
notions became dubious, and by the middle of the 19th century it had become clear that life never arises anew. Every living thing descends from some
previous living thing, and spontaneous generation is a fantasy. The proposition that the early earth was lifeless and that life arose at some point in time

flatly contradicts the dogma. Biologists are tempted to sidestep it, and focus on more concrete matters that we know how to handle.

But the origin of life is also the most consequential issue in biology, perhaps in all of science. Until we understand how life came into existence, we cannot rigorously exclude the possibility that it required an act of creation by powers that go beyond natural processes. Only when that mystery has been resolved will it be clear how life is connected to the nonliving universe, and only then will biology be securely anchored in the physical sciences. The fact that after some eighty years of debate and experimentation we still do not have a convincing theory of the origin of life is disconcerting, and downright embarrassing. The object of this chapter is to summarize what has been learned, and what remains to be discovered.

The Toughest Nut of All

However life came into being it happened an almost inconceivably long time ago, when the planet was still in the process of formation and a very different place from today. The most ancient traces of microbial life go back at least 3.7 billion years, suggesting that the earth has hosted life for roughly 4 billion years. Until about 4.2 billion years ago the earth was roiled by violent turmoil, including the event that threw off the moon (about 4.5 billion years ago), and intense meteorite bombardment that could have set the seas a-boil and sterilized the surface. If so, life will have emerged during a window of a few hundred million years, presumably in some sheltered locale. During that time the atmosphere was probably dominated by nitrogen and carbon dioxide, and entirely devoid of oxygen. Early life must have been strictly anaerobic. The seas were probably salty, as they are today. Whether life arose in salt water or in fresh is under debate, but there is some evidence that the water was very warm. However life began, it must have been a hardy plant that could take hold and flourish under unpropitious circumstances. And the fact that all living things are fundamentally of the same kind suggests that the origin of life was in some sense a singular event. A singular creation? Not necessarily; it may be that all contemporary organisms are descended from the few survivors of some early cataclysm. In any case, we must accommodate an early bottleneck.

One basic question is whether life originated on earth or traveled here from elsewhere. The latter notion, first proposed more than a century ago, is presently quite out of fashion but remains in the running. It is surely conceivable that "seeds" of life (bacterial spores, perhaps) hitched a ride inside meteoritic rocks, or were planted by interstellar voyagers, accidentally or on purpose. There is, of course, not a shred of evidence for "panspermia," and it is most unlikely that relevant traces could have endured for 4 billion years. But the hypothesis is in principle testable, and may become so should we ever encounter life beyond earth and be in a position to examine its chemistry. If it turns out that alien life shares the molecular specifics of the terrestrial version (e.g., DNA, RNA, proteins made of amino acids), the possibility of interstellar transfer will have to be seriously considered. For the present, however, the great majority of scientists are content to set it aside, and so shall I.

Assuming, then, that life arose from lifeless matter here on earth, we must consider whether this can be wholly accounted for by natural causes or required an act of creation by some sort of advanced or transcendent intelligence. It is not possible to exclude God's finger from the inception of life, but if so He left no trace of His intervention. In the absence of such evidence the postulate of "intelligent design" by an extraterrestrial agency takes the origin of life clear out of the realm of science, and effectively terminates the inquiry. The same is true of the hypothesis that life began with some exceedingly rare chance event, effectively a miracle (that, incidentally, was the view of Jacques Monod, whom I quoted in the epigraph to Chapter 1). I can imagine observations that would compel us to take the idea seriously, but for the present anyone committed to the quest for a rational understanding of the world must set it aside as well.

In practice, the only hypothesis that keeps the origin of life within the purview of science is that life arose on earth by natural causes alone. That implies that there is nothing unique or miraculous about genesis; it would probably happen wherever in the universe conditions are favorable. That is the position adopted as the starting point by almost all scientists, myself included, but we must not deny that it bristles with difficulties. Science does best with repeatable, law-governed processes and events; a rare one such as the origin of life is hard to apprehend. Indeed, it is not at all clear what observations could falsify the claim of spontaneous generation billions of years ago, and therefore whether we are still on science's playing field at all. All the same, the idea has garnered a body of evidence that, while falling well short of proof, is consistent with the viewpoint of a natural genesis. We do not presently understand how life began, or where, or exactly when. But we can be reasonably confident that chemistry and physics do not forbid a natural genesis, and that its pursuit is a legitimate scientific inquiry.[2]

Biology's Black Hole

The first legible entry in the long annals of life is LUCA, the Last Universal Common Ancestor. The evidence supporting the claim that all living things shared a common ancestry about 4 billion years ago is necessarily indirect but persuasive (Chapter 6), and supplies a solid base for reflection on life's subsequent history. Technically, LUCA is the first node on the universal tree of life (Figure 6.1), the time when Bacteria and Archaea diverged, and it is far from certain what kind of creature that point represents. Common ancestry does not necessarily mean that there was a specific common ancestor. There is reason to suspect that LUCA stands for a stage in the evolution of cells, a community of precellular entities that readily swapped genes, and lacked the discrete lineages that we associate with organisms as we know them today. But if we have read the clues correctly, LUCA will already have been an organism of sorts: a complex system made up of molecular elements specified by genes, that drew matter and energy into itself, maintained its identity, reproduced, and evolved. If we could inspect a specimen we would probably recognize LUCA as living, or at least well on the way.

So where do we go from here? LUCA was not really primitive; she must herself have been the product of a lengthy evolutionary history, whose agents were yet more shadowy entities loosely designated "protocells." It is the protocellular phase that will have generated all the features that make life possible: metabolism and membranes, devices to harvest energy and put it to work, proteins, nucleic acids, genes, ribosomes, and the elaborate machinery by which a sequence of nucleotides spells a sequence of amino acids and specifies a function. More than that, the protocellular phase must have generated the first organized structures, be they cells or their immediate precursors.

The later steps in the shaping of LUCA can be attributed to standard evolutionary processes. One can tell, for example, that ribosomes and the genetic code were optimized by mutation and selection, and that certain protein families formed and diversified even before LUCA. The question is, how could life ever get to the stage where concepts such as heredity and physiological functions have their contemporary meanings? Evolution by variation, competition, natural selection, and adaptation can only kick in once a substantial degree of functional organization has been attained. Unfortunately, the advent of the standard mechanisms for the expression and replication of genetic information (Chapter 3) brought down an opaque curtain which hides all that came before. What did evolution look like before these underpinnings were in place?

Here is the black hole at the root of biology. If the origin of life is to be kept within the purview of science, we must envisage its emergence by natural causes from the dust of a planet barely out of its infancy. Scientists assert that life arose from lifeless chemistry or geochemistry, and then proliferated to smother the earth in a multitude of creatures while the world kept changing relentlessly all around. Chemists puzzle over the origin of the sophisticated molecules that support the activities of living things, all so highly interwoven that one cannot see how any one could have arisen in the absence of the others. Biologists ask how organisms came to be, how purposeful behavior could have emerged from of an inorganic setting that gives no hint of such capacity. There is no model, no analog, and no precedent to lean on. On the face of it, a spontaneous genesis is a ludicrously implausible proposition, the worst possible hypothesis—except for all the alternatives. No wonder that we are making heavy weather of the beginnings of life, nor that most practicing scientists would rather think about something else.

It is essential to appreciate that the origin of life is not solely or even primarily about the invention of its remarkable molecular framework. Biomolecules by themselves are inert; only when they are organized into a cell do molecules come to life. The vital question is, whence came the first chemical systems that had the capacity to grow, reproduce, and evolve? This quest is quite alien to the reductionist spirit that underpins the brilliant achievements of contemporary biology. Instead of dissecting existing structures and operations into their constituent parts in order to discover how they work, we now seek answers to a historical question: how working systems came to exist and acquire the elementary parts required for a functional whole.

Selected for Dynamic Stability

The preceding ruminations, however inconclusive, have at least sharpened the issues. We need to discover how life could have begun in chemistry or geochemistry, and how it persisted and proliferated in the face of constant change. These are historical questions, matters of when, where, and how, but perhaps not entirely historical. If indeed life is a cosmic phenomenon that is likely to spring forth wherever and whenever conditions are favorable, then beneath the many contingent factors there ought to be a stratum of necessity, natural processes that we can discern and perhaps probe by experiment. The discussion that follows draws heavily on the writings of the Israeli chemist Addy Pross, who has brought clarity and fresh insight to a conversation that had gone stale.[3]

If life began in chemistry it will have started simple and grown increasingly complex and autonomous over time. Contrary to what creationists sometimes claim, this assertion does not violate the laws of physics as long as mounting organization is paid for by energy drawn from the environment; but it does contradict the commonsense understanding of how the world works. We cannot call on Darwin's ratchet of heredity, variation, and natural selection, because at the onset of life the requisite framework of organized molecules was not yet in place. Can a purely chemical system be subject to variation and natural selection, and why would it evolve complexity, integration, and functions to reach a level that we recognize as "living"? Selection, alright, but for what quality? Pross argues (as have others) that in the beginning chemical systems were selected for stability, the capacity to persist, and that the ceaseless search for stability is the force that underlies all of biological evolution.

Nature displays two very different kinds of stability. The familiar kind is that of solid, stolid rock, which persists because it interacts very little with its surroundings.

The other and more interesting kind is dynamic kinetic stability, illustrated by the bicycle that stays upright as long as it keeps moving—courtesy of someone pumping the pedals. The stability of living things is like that: they draw matter and energy from their environment and persist thanks to this ceaseless commerce with the world beyond their boundaries. That is the reason living things are unavoidably complex: their very persistence requires them to perform a multitude of tasks, to metabolize, renew all their constituents, and to assemble their own structure. Dynamic kinetic stability is uncommon in nature, but it works for living things: individual creatures are mortal and transient, but the pattern of biological organization that each individual represents can outlast the mountains.

If dynamic stability is one essential aspect of the emergence of life, reproduction is another. Dynamic systems are subject to wear and tear that will eventually destroy them. Living organisms evade the inevitable by reproducing themselves before the rot sets in. Biological reproduction is the output of an exceedingly sophisticated procedure, but embedded deep within it is a much simpler chemical principle: the replication of a nucleic acid, most commonly DNA. The key role of replication has been recognized for decades, and molecular scientists are apt to take it as an article of faith that the road to life began with the spontaneous appearance of a self-replicating molecule, most likely RNA. Whether or not that is the case, there is no doubt that the incorporation of such a molecule in a dynamic system will radically transform its prospects, for entities that reproduce themselves quickly outrun everything else. Natural selection of a self-reproducing system for stability (persistence) should be rewarded by the chance acquisition of ancillary devices, especially the capacity to harness energy. The system will tend to become more complex, with ever more moving parts, and now we are launched on the road to life.

It seems that evolution in the Darwinian manner requires some kind of self-reproduction at its core. This insight does not explain how useful functions such as genes or energy transformation came into existence, but it does assure us that, once discovered, they will be drawn into the service of a self-reproducing system. Each individual system is mortal, but by reproducing its pattern again and again the great fountain of life can continue to play indefinitely. Reproduction is not an optional feature: like dynamic stability and complexity it is part of life's very essence.

In the beginning, then, there will have been only chemistry: perhaps a small network of chemical reactions resulting in the reproduction of a replicator at its core. As Pross puts it, we should not think of replication as a manifestation of life; on the contrary, "life is a manifestation of replication."[4] The argument that reproduction supplies the driving force for the elaboration and proliferation of life seems to me an important insight, and I shall adopt it in what follows. Still, serious doubts persist as to how that could have begun: The spontaneous appearance of a self-replicating molecule would not be sufficient to spark life. Replication requires, at the least, precursors primed for assembly and a local environment conducive to that kind of chemistry. Almost certainly we must postulate a basic system from the start, endowed with some degree of organization, whose origin is not obvious. It would be reassuring if we had a model, any kind of system that reproduces itself in the absence of biological catalysts and clever chemists!

Ways and Means

In the preceding section we sought insight, not into How life might have come into existence, but Why: What force could drive ordinary chemical reactions to grow into complex, integrated networks that can make and perpetuate themselves? Addy Pross, whose lead I followed, finds an answer in the tendency of all systems to change from a less stable state to one that is more stable. In the case of open dynamic systems, the stability they seek is not static but kinetic: Open systems persist, not in spite of incessant interactions with their environment but because of them. This universal tendency will be powerfully reinforced if the chemical system in question brings about its own reproduction; and will be further stimulated by the acquisition of ancillary components, particularly a supply of energy. Pross draws his thesis from the most general principles. It should apply whenever and wherever a self-reproducing network has arisen, and is compatible with a range of different chemistries. It makes genesis a manifestation of chemical laws rather than the product of mysterious or miraculous happenings, and that should appeal to anyone who seeks a rational understanding of nature.

Our focus in the present section is rather more earthly. Given that the only kind of life we know is the terrestrial version, what can one surmise regarding the chemical system that spawned it, and the physical circumstances under which that occurred? Keep in mind that we are speaking here of a time before genes, functions, organized structures, and the Darwinian mode of evolution. With the actual prehistory of life inaccessible, scientists have rephrased the question to ask how life, any life, could have come into existence. The current objective is to construct a simulacrum of a living cell in the laboratory, using the molecular components of contemporary cells. Despite this limitation, the gigantic literature makes a rich trove of data for reflection on the possible course of events that resulted in the rise of our sort of life.

Soup for Starters?

Was life based on carbon from the beginning, or did it start in the geochemistry of minerals? Life today relies on a small subset of elements (carbon, hydrogen, nitrogen, oxygen, phosphorus, and sulfur, or CHNOPS) plus lesser amounts of others. Molecular scientists mostly take it for granted that this will have been true from the beginning, but a small school of dissenters argues that life must have begun with the geochemistry of minerals and switched to an organic basis in the course of evolution. Like most of my colleagues I find that proposal hard to believe, but geochemistry will make a comeback in the context of life's first energy source.

The most popular scenario traces the first constituents of life to a "primordial soup," a rich stew of organic substances produced by various nonbiological reactions, that would have accumulated on the young earth because there were no living organisms to consume the products. In principle, the first organisms might have assembled themselves from a menu of prefabricated components, conceivably even including proteins and nucleic acids. There is, of course, no direct evidence, but the proposal that some sort of prebiotic broth existed is surely plausible. It has spawned a small but vigorous industry devoted to "prebiotic chemistry": the systematic search for procedures to generate all sorts of biological molecules under conditions that might have prevailed on the young earth. Some practitioners even hope to reconstitute life in the laboratory, with such abiotic molecules as building blocks.[5]

Skeptics suspect that the scenario outlined thus far is a little too convenient to be true. Simple molecules of abiotic origin may well have been present and found a role early on, but self-assembly ignores the need for energy and fails to connect with both evolution and contemporary metabolic pathways. If we begin from the proposal that life began with a simple dynamic system seeking greater kinetic stability, most of the biomolecules (and especially the information-bearing proteins and nucleic acids) will have evolved concurrently with protocells and entailed the progressive invention of metabolism. The significance of prebiotic chemistry then resides not in its products as such but in the delineation of chemical mechanisms that protocells could have found and put to use.

Enclosure

Until quite recently it was generally thought that the central pathways of metabolism, and even the apparatus of protein synthesis, first arose in free solution. That position never made sense, and has largely been abandoned. Today's conventional wisdom holds that enclosure of some kind was necessary from the beginning, if only to keep the reactants together and to maintain an environment conducive to biological syntheses. Contemporary cells accomplish this with lipid bilayer membranes, and most experimentalists look to lipid vesicles of abiotic origin to get things started.[6] Membrane-forming lipids will likely have been present in the soup, though not the phospholipids that are ubiquitous in contemporary organisms. The trouble is that membranes tight enough to perform their expected function will also have been highly impermeable, and would have to rely on special portals to pass nutrients and ions from one side to the other. Where would those have come from?

Whence Came the Energy?

Most of what living things do, including all forms of biosynthesis and re-production, come under the heading of "work": they will not take place without an input of energy. That must have been true from the beginning, but how protocells (and LUCA, for that matter) captured energy from the environment and put it to work remains entirely unknown. Contemporary organisms universally accomplish this with the aid of an ionic current across a membrane, usually carried by protons, and make use of an ion-translocating ATPase (Chapter 3). This procedure is thought to be extremely ancient, possibly going back to the earliest systems, which again points to enclosure in a membrane studded with transport carriers. How, prior to the invention of proteins and complex membranes, can one imagine the acquisition and utilization of energy?

This dilemma has drawn attention to submarine hydrothermal vents, particularly the subclass of vents that are warm rather than scalding and alkaline rather than acidic. These vents are places where geochemical effluents, laden with hydrogen gas and methane, interface with ocean water bearing CO_2. They lay down massive spongy deposits, whose nooks and crannies could provide secluded chambers where products of CO_2 reduction could accumulate and undergo further transformation. Potentially, such vents may have been the cradles of life. Even today, there are Bacteria and Archaea that make a living from the reduction of CO_2, and these may be drawing on metabolic pathways that first evolved with the earliest cells. It has even been suggested that the gradient of pH between alkaline effluents and more acidic ocean water served as the first source of protobiological energy, and the precursor to today's sophisticated mechanisms of membrane-linked energetics.[7]

The idea is romantic and intriguing, but like all other proposals for the origin of life it is beset with difficulties. The postulate that the long history of life began with the reduction of CO_2 by geochemical hydrogen gas catalyzed by inorganic minerals stuck in many a craw, mine included. True, the reaction is thermodynamically feasible, and the microbes that live by it appear to represent very ancient lineages. But it calls for elaborate biochemistry that looks anything but primitive, and it underpins existence on a tight energy budget. Assimilation of CO_2 as it operates in extant organisms is surely the product of prolonged adaptation. But just in the past two years a small spate of papers has supplied evidence for a particular variant of the geochemical hypothesis:[8] in the presence of iron as catalyst, CO_2 can be reduced to pyruvate, which goes on to generate a set of compounds familiar from a major metabolic pathway, the (reversed) tricarboxylic acid cycle. Both high pressure and a gradient of pH facilitate the reaction. As always, the devil lurks in the details; and it is too early to judge where this will lead. Still, all of a sudden what had been an imaginative but somewhat marginal idea basks in real evidence to buttress its links to both geochemistry and contemporary metabolism.

Genes and Functions

If one wished to single out one feature to represent the secret of life, it would surely be the interplay of DNA, RNA, and protein; or in more biological language, heredity, reproduction, and function. We understand quite well how the procedures work, but how they could have come into existence continues to defy the imagination. Biochemists note that neither DNA nor proteins can be made in the absence of the other, and that it also requires a lot of clever machinery. Organismic biologists wonder how a sequence of nucleotides came to represent a function useful to a larger entity, such as an enzyme. Of all the unanswered questions in cell evolution this is by far the most wicked, so much so that it raises doubts whether evolution, beginning in systems chemistry and proceeding by small steps, could really achieve so spectacular an outcome.

An influential proposal to break the impasse began to emerge early during the rise of molecular biology, subsumed under the heading "RNA World."[9] The central idea is that in the beginning were neither DNA nor proteins, and that the roles these perform today were played by RNA. The chemical structure of RNA allows it to carry sequence information as genes do, and also to catalyze chemical reactions as proteins do. The idea that initially everything was done by RNA alone was dramatically reinforced in the 1980s by the discovery that several important cellular function are carried out today, not by the usual proteins but by "ribozymes," catalytic RNA. One of these functions is protein synthesis by ribosomes: the critical reaction that links one amino acid to another is performed by ribosomal RNA (yes, the same RNA that serves as the chronometer in molecular phylogeny, Chapter 6). Proponents of the "RNA World" came to hold that a ribozyme capable of replicating its own structure arose spontaneously in the primordial soup, acquired ancillary molecules (RNA and prebiotic peptides), and launched the biotic enterprise. The proposal has undergone much revision, but the principle continues to resonate: the hypothesis adopted here, that life began with an autocatalytic system of reactions capable of reproducing itself, is one of its offspring.

The appeal of an RNA-centered beginning set off an intense effort to demonstrate experimentally that RNA can replicate all by itself, in the absence of protein enzymes. The search for a simple replicator has not been successful but more complex systems work well, such as two RNA species replicating together, each catalyzing the synthesis of the other.[10] This strongly supports the recognition that replication will have been from the start the job of a network of reactions that would also supply the requisite precursors, primed for assembly, and a sequestered nook to work in. Just how such a basic network would have come into being is now the burning question.

Fast forward to the next stage, and we come up against that whole sophisticated machinery of translation, both biochemical and semantic. How did it come about that a particular sequence of nucleotides corresponds to a cognate sequence of amino acids, and also performs a function useful to the stability or reproduction of a larger entity? There is an immense gap between self-reproduction of an RNA sequence and a set of instructions that programs a function, and no such transition has ever been seen in nature. Researchers have sketched out some ideas, but none that triggers the "Eureka!" reaction. We are far from done, and just how chemistry morphed into biology continues to pass understanding.

Something Missing

A credible path to the origin of life remains to be discovered. This is an active field of research, spurred by happy talk both among scientists and in the popular press; my own take is considerably more skeptical. As I read the literature, it seems plausible that the road to life began with a small network of chemical reactions that had the capacity to reproduce itself. At its core may have been some kind of self-replicating molecule (possibly but not necessarily RNA), and it would have included ancillary reactions to supply energy and some degree of structural organization. Evolution in the direction of greater autonomy, complexity, and functional organization would have been driven, not by the classical interplay of heredity, variation, competition, and natural selection, but by the search for increased dynamic kinetic stability. The exact nature of this system and where it may have come from are entirely unknown and represent the current version of biology's black hole. However, it is not unreasonable to hope that future research into the chemistry of replicating systems will turn up suggestive examples.[11]

Granted that such a rudiment of biological organization had somehow come to exist, there remains a huge gap between a chemical network and even the simplest protocell; and it is not at all clear how (and even whether) evolution by small steps could bridge it. The widest of the chasms to be leapt would have been the invention of translation, the production of proteins that perform a cellular function at the behest of instructions spelled out in a nucleotide sequence. To my knowledge, no plausible scheme to accomplish this feat has yet been proposed, and there is no analog or precedent. To paraphrase a remark by J. T. Trevors and L. D. Abel,[12] natural processes have never been observed to write prescriptive instructions. Protocells, beginning with nothing more than a prebiotic soup, must be credited with an achievement that has so far eluded the best and brightest of chemists. One cannot help wondering whether we have been missing something fundamental.

One aspect of the problem that has received remarkably little attention from mainstream investigators, intensely preoccupied as they are with the RNA world, is the place of energy. A source of energy, and some machinery to harness it, must have been part and parcel of the origin of life, for two reasons. First, energy will have been required for the production, collection, and activation of the components of that (hypothetical) seminal self-reproducing network. Second, a flux of energy is a prominent source of spatial order in dynamic systems. Forty years ago, Ilya Prigogine drew much attention to the "dissipative structures" that regularly arise in physical systems kept off equilibrium by energy flow. Familiar examples include flames, hurricanes, the vortex created when water flows out of a bathtub, and those remarkable "Benard cells" produced when a shallow pan of liquid is heated from below. For a time many biologists thought that dissipative structures would cast light on the production of biological forms. That hope proved misplaced—morphogenesis is better understood as the outcome of self-organization by gene-specified elements, not of energy flux (Chapter 3). But before there were genes and proteins, only a network of chemical reactions, a dissipative structure may have initiated the rise of spatial organization.

There would have been no shortage of energy impinging on the (hypo-thetical) primordial broth, including light and lightning, but nothing that meshes in a suggestive way with contemporary biochemistry. That is why I will not give up on those submarine hydrothermal vents, where gradients of redox potential and pH exist naturally and sustain themselves for tens of thousands of years. For all the difficulties in discerning a plausible coupling mechanism, I see no better prospect for a primordial source of biologically relevant energy. Geochemical energy is not a magic bullet—it won't help solve the problem of translation—but it may be the best way to get things started.

Scientists, myself included, cling firmly to the premise that life arose here on earth by natural causes, without help from transcendent powers, extraterrestrial visitors or wildly improbable chance events; and we have good reason to do so. One is the long-standing principle of rational inquiry to exclude a priori anything that is not accessible to reason. The other is that none of the alternatives offer a productive explanation for how life came to exist; they do little more than put the entire matter out of reach. Now, such arguments do not necessarily make radical alternatives untrue, and given the persistent intractability of the origin of life nothing can be taken off the table. I cannot shake off the nagging hunch that there is more, much more, to the question of origins than is dreamt of in contemporary scientific discourse. If and when someone devises a persuasive scenario for how life came to be, it may be situated in a world radically different from the one we currently take for granted.

PART III
THE GYRE OF COMPLEXITY

8

The Expansion of Life

I like to compare evolution to the weaving of a great tapestry. The strong unyielding warp of this tapestry is formed by the essential nature of elementary non-living matter, and by the way this matter has been brought together in the evolution of our planet. In building this warp the second law of thermodynamics has played a predominant role. The multi-colored woof which forms the details of the tapestry I like to think of as having been woven onto the warp principally by mutation and natural selection. While the warp establishes the dimensions and supports the whole, it is the woof that most intrigues the esthetic sense of the student of organic evolution, showing as it does the beauty and variety of fitness of organisms to their environment.

—Harold F. Blum, *Time's Arrow and Evolution*[1]

As success stories go, nothing matches the history of life. Current accounts begin about 4 billion years ago with something altogether novel, a set of chemical reactions that could persist and reproduce itself. In time this system evolved and expanded to blanket the entire planet with organized, purposeful biochemistry. The expansion of life was untidy, irregular, and episodic but relentless, and it produced a vast tide of living things of stunning diversity. If variety is one side of the biological coin, unity is the other: all living things on earth are of one kind, members of a single biochemical family, and all are made of cells (one or many) which are the fundamental units of life. In some sense, the advent of life seems to have been a singular event. We can only speculate whether what happened on earth was unique, or but one instance of a phenomenon widespread in the cosmos. This chapter will examine the burgeoning of life in the broadest terms, looking both at how it came about and why.

At bottom life is a microbial phenomenon, more specifically a prokaryotic one, and the first 2 billion years of life's history feature nothing but prokaryotes. It is they and their forerunners that discovered how to build macromolecules and cells, to harvest energy, and to manage genetic information. This is why the preceding chapters focused largely on microbes, with scarcely a mention of higher organisms. But the living world today is far richer, dominated by plants, animals, and fungi on a scale of size thousands or millions of times larger and composed of billions of cellular units; and it is these that will hold center stage in the remainder of this book. Prokaryotes did not go extinct, far from it, and in many respects it's still a prokaryotic world. But I take it as a basic premise that the expansion of life revolves around increases in size, complexity, structural organization, and functional capacities, and that in some elusive sense, evolution is "progressive."

Three general features of life's expansion should be underscored from the outset. First, all higher organisms are made up of eukaryotic cells. Second, all of them are multicellular. It is true that prokaryotes went some way in the direction of higher-order organization (the multicellular fruiting bodies of mycobacteria are a case in point), but they always fell short. It is also true that certain unicellular eukaryotes, ciliated protozoa in particular, boast astonishingly complex cells and achieve remarkable feats. That a small fragment of the ciliated protozoan *Stentor*, less than one hundredth of the cell, can regenerate an entire organism never ceases to astonish me. Still, the architectural complexity of the ciliates does not match that of plants or animals, nor do their physical and mental abilities. There evidently is a limit to what is possible within the confines of a single cell, and this barrier can be overcome by harnessing many individual cells to a common purpose. The third general feature of life's proliferation is that it came about by the interaction of many kinds of organisms that compete, consume each other, cooperate, or just coexist. What has proliferated so spectacularly is not some particular kind of organism, but the phenomenon of life.

In the great theme of life's expansion, the processes that sustain life matter more than the particular organisms that embody life. Life's history is like a play, whose plot guides what the characters do (or, as Ford Doolittle has put it, it's about the song rather than its singers). More often than not, the plot calls for organisms of different kinds working together. Organisms are forever exploring fresh niches on the surface of others or inside them, and one kind exploits another's waste. Life is not a passive respondent to its environment, but actively shapes and constructs habitats. Complexity breeds greater complexity, generating a gyre of mounting functional organization. This point of view offers a perspective complementary to the more conventional celebration of showy organisms and "missing links," one that highlights the inventions, turning points, and transitions that confer a degree of direction on the succession of events.

Synopsis of the Plot

The scientific version of genesis, as we presently understand it, unfolds in four movements (Figures 6.2 and 8.1). It begins with a prologue, set about 4 billion years ago, which tells of the origin of life and the evolution of cells. What transpired during that time is not well understood (Chapters 6 and 7), but there is very good reason to believe that it culminates with the appearance of the contemporary kind of cellular organization. All organisms that exist today trace back to a singular, hypothetical entity called LUCA, the Last Universal Common (or Cellular) Ancestor. Note that all the essential features of life made their appearance during that initial 5% of biological history, which remains largely inaccessible to us.

The oldest known fossils, about 3.7 billion years old, mark the beginning of life's first era, one in which prokaryotic microorganisms (both Bacteria and Archaea) made up all the biosphere. Beginning with a sparse population of pioneers that may have relied on geochemical sources of energy, prokaryotes flourished, diversified, and gave rise to all or most of the phyla known today. They progressively discovered all the pathways that access energy, most significantly photosynthesis and especially that mode of photosynthesis in which water serves as the reductant and oxygen is released as a byproduct. For most of that era oxygen was absent or very scarce, but around 2.3 billion years ago enough oxygen accumulated to turn the atmosphere oxidizing, and the world changed for evermore.

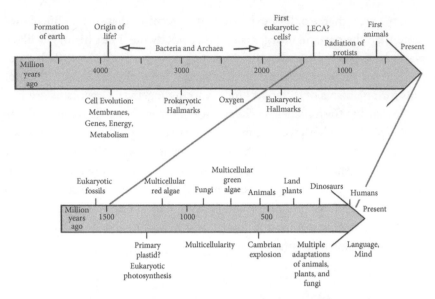

Figure 8.1 Timeline for the expansion of life. Organisms are listed above the bar, transitions and innovations below.

The second phase of biological evolution, the coming of the eukaryotes (Figures 8.1 and 8.2), is marked by the appearance in the fossil record of relatively large but enigmatic objects generally taken to be fossils of the first eukaryotic microbes. The oldest date back to about 1.6 billion years ago, and appear to represent lineages only distantly related to contemporary protists. Just what it is about the eukaryotic order that allowed them, but not prokaryotes, to go on to multicellularity is not entirely clear. I am impressed by the argument that eukaryotic cells command more energy per gene than prokaryotes do (Chapter 6), but others remain skeptical.

The origin of the eukaryotic cell, a transition second only to the origin of life itself, remains intensely controversial (Chapter 6); but one aspect has been established beyond reasonable doubt: mitochondria are descendants of Bacteria that took up residence in the cytoplasm of a host cell. The nature of that host is uncertain. Current opinion, which I share, holds that it was most likely an early archaeon, and that the acquisition of those endosymbionts was the crucial step in the evolution of the eukaryotic order. If that is correct, eukaryotic cells are products of a very rare, possibly unique event. Subsequent evolution entailed the invention of a long list of novel processes and

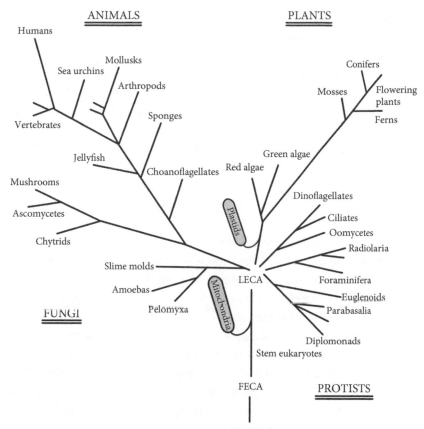

Figure 8.2 The advance of multicellular life. The diagram highlights only the three most spectacular instances of multicellularity (animals, plants and fungi), omitting many more minor ones.

organelles (mitochondria, nuclei, and mitosis, cytoskeleton, cilia, etc.), producing cells larger and much more elaborately organized than prokaryotes.

All eukaryotes living today appear to share a common ancestry. LECA, the (hypothetical) last eukaryotic common ancestor, may have lived about 1.3 billion years ago (with a wide margin of error) and featured all the hallmarks of contemporary protists, including motility, phagocytosis, and sexual reproduction. LECA sits at the base of a great radiation that produced the cornucopia of contemporary protists, many of which are thought to be comparatively recent. Amoebas and diatoms, for example, first show up about a billion years ago or less.[2] Many protists live by photosynthesis, thanks to another major transition: In a second instance of endosymbiosis an early eukaryotic protist took up a cyanobacterium, and eventually acquired the capacity for oxygen-producing photosynthesis (the timing is disputed). Almost all contemporary plastids descend from that primary endosymbiotic partnership, but a few owe their origin to another, later and independent event. Subsequent secondary endosymbioses spread photosynthesis around the eukaryotic universe. It is among the protists that we seek the roots of life's third stage, the rise of multicellular (higher) organisms.

The flowering of eukaryotic life took place in the context of an earth continuously changing in response to both geological and biological processes. The one that has caught the eye is the level of oxygen in the atmosphere, itself a product of microbial photosynthesis.[3] The evidence indicates the oxygen accumulated in two discrete steps. First, during the prokaryotic era, it rose with fluctuations from near zero to reach 1% or less of today's level about 2.3 billion years ago. That is the time when the atmosphere turned mildly oxidizing. That landmark was followed, about 550 million years ago, by a rapid rise to near contemporary levels. Just how and why oxygen increased at that time is not altogether clear, but it will have been a consequence of both biological and geological forces. The level of oxygen represents the balance between processes that raise it (primarily photosynthesis) and others that lower it, notably respiration and weathering. A net increase requires the sequestration of some of the organic matter produced by photosynthesis, thus shielding it from the attentions of microorganisms. The deposition of coal and petroleum are instances of carbon burial, but just what accounts for oxygen accumulation 550 million years ago is not certain.

It has not escaped notice that the transitions in oxygen level are correlated in time with major biological ones: first the rise of the eukaryotes beginning some 2 billion years ago, and then the rise of animals 550 million years ago. Is there a causal connection? It seems very likely. Eukaryotic cells are fundamentally aerobic, dependent on several biochemical processes that require oxygen; not to mention the key role that mitochondria seem to have played in the very origin of that mode of cell organization. As to animal life, the active mobile lifestyle so characteristic of animals cries out for a plentiful supply of energy. The suggestion that animals had to await a sufficient supply of oxygen is reinforced by the discovery that extra-high levels of oxygen during the carboniferous coincides with an efflorescence of insect life. Fossil dragonflies with a wingspan of a foot make the point.

Should we then conclude that the rise in oxygen level *caused*, first the advent of eukaryotes and then of animals? No, but it helped establish conditions that life could exploit in the service of its own inherent drive toward increasing complexity and organization.[4] The same could be said of other environmental changes, including the shifting of continental plates and periodic global glaciations. All will have contributed to the physical warp that holds together the tapestry of life, allowing variation and natural selection to shape the details.

The proliferation of life during the most recent billion years revolves around the advance of multicellular organisms, which now make up more than half the world's biomass: colonial algae, massive oak trees, salmon and minnows, and fungal colonies many yards in diameter. By definition, in a multicellular organism cells interact, communicate, form specialized organs, and submerge their individuality in a larger whole. Thanks to the rich fossil record archived in sedimentary rocks, we are well informed about this third stage in biological history, and most readers will be familiar with its outlines. Just for a reminder, the first animals that are clearly multicellular show up in Precambrian strata: tube worms, snails, tracks, and burrows.[5] Their most conspicuous representatives make up the Ediacaran fauna that flourished around the world between 635 and 540 million years ago. These creatures lived in shallow seas and typically had large flat frond-like bodies as much as a foot in length; they lacked not only limbs but also a mouth and an anus, and how they made a living remains a mystery. The Ediacarans were completely displaced in one of the most striking episodes in biological history, the "Cambrian Explosion," which saw the almost simultaneous appearance of all the animal body plans that we still recognize today. With the Cambrian began the tremendous radiation of animals, plants, and fungi that made the living world as we know it.

The counterpoint to this progressive reading of life's history is extinction. The vast majority of all organisms that have ever lived are extinct, their places in the ecological theater taken by others that may, or may not, be descendants of the prior inhabitants. The unfolding of the living world is slashed by recurrent episodes of mass extinction, some of which wiped out a large proportion of higher organisms. The meteorite that slammed into the earth about 65 million years ago and is credited with bringing to an end the rule of the dinosaurs is probably the most familiar instance, but an even greater mass extinction decimated the living world at the end of the Permian era, 252 million years ago. Extinctions reshuffle the deck; we do not know whether those who survived had the right genes, or merely better luck.

The advance of the multicellulars brought numerous spectacular novelties and adaptations, but no fundamentally new kinds of organisms have appeared during the past 400 million years. The one possible harbinger of a radically different fourth phase of evolution may be the advent of humans, and of mind. Whether or not that proves true remains very much in the balance, perhaps the most fateful question that we can ponder.[6]

Going with the Flow of Energy

Earth sits athwart a torrent of energy that flows from the sun to the vast sink of outer space. Of the trillions of calories that reach the surface some are reflected, some are absorbed and radiated again as heat. Less than 2% of the energy is picked up by photosynthetic organisms, but that modest share powers the biosphere: either directly, as in plants, algae, and some bacteria, or indirectly as in the animals, fungi, and bacteria that consume the primary producers. The only exceptions are those microorganisms that rely on geochemical sources of energy, such as the reduction of CO_2 or ferric iron by hydrogen gas. We have traditionally thought of these as a minor wrinkle, but that may be incorrect: if certain current musings are correct, the "deep hot biosphere" is as large as the one at the surface,[7] and the very origin of life was fueled by geochemistry.

What do organisms do with their energy harvest? Almost everything. We saw in Chapter 3 that most of what they do comes under the heading of work: biosynthesis, movement, maintenance, growth, and reproduction can only take place thanks to the input of energy. "Living things" are not things at all but processes, powered by the continuous consumption of energy. Broadly speaking we understand how this energy is captured, converted into chemical forms that organisms can use, and put to work. And we recognize that these crucial operations rely on a small set of chemical processes that are largely shared by all living things (Chapter 3). In this section we shall examine a less conventional aspect of the biological energy economy: the proposition that the flux of energy not only sustains life from day to day but also constitutes the force that drives the expansion of life and ultimately underlies all of biological organization. This is not a standard hypothesis subject to verification and possibly falsification; instead, it is a point of view that illuminates the connection between evolution and its physical context.[8]

The argument runs like this. Events in the physical world are governed by the second law of thermodynamics, which mandates that the general tendency of things is to fall apart. The ultimate destiny of all order is to dissipate into chaos. Living things postpone the inevitable by doing work to repair the ravages of time, restore gradients, and hold out against entropy; and they have done so on the grand scale, endured for four billion years, and flourished prodigiously. What makes this possible is that the second law has two faces: the familiar destructive one, and a more cryptic constructive one. Whenever the dissipation of energy is constrained it "seeks" new paths by which to flow, and these commonly entail the construction of orderly systems that accelerate the downhill rush.

The capacity of energy flux to generate order, at least locally and transiently, is well documented in the realm of chemistry and physics. Ultraviolet light falling on certain mixtures induces cyclic chemical reactions that dissipate the energy gradient. Flames, vortices, hurricanes, and also those astonishing hexagonal flow-structures called Benard cells are all physical structures directly shaped and maintained by the flow of energy from a higher level to a lower one; and the structures that the flow builds themselves accelerate the dissipation of energy. The term "dissipative structures" aptly describes their nature. What about living organisms? They too depend on the continuous flux of energy to construct, maintain, preserve, and reproduce themselves, but living things are not dissipative structures. They are shaped by internal and largely autonomous forces, not by energy flow. In living organisms, the role of energy is permissive rather than instructive: it sets the boundaries within which life must tack to find a course compatible with its own internal imperatives. But the propensity of energy flow to generate order becomes apparent at a higher level, that of ecosystems.

Imagine a meadow in summer bursting with life, glad with flowers, and buzzing with bees. Amid all the buzzin' bloomin' confusion, order does not leap to the eye, but ecologists have realized for nearly a century that ecosystems are structured by the flow of energy. In the meadow, incident light is first absorbed by green plants and put to work. Plants support smaller populations of herbivores, from snails and butterflies to sheep, and these in turn sustain still smaller populations of carnivores—dragonflies, coyotes, and shepherds. At each one of these "trophic levels" energy drains away as its bearers die, in the process sustaining decomposers such as beetles, fungi, and bacteria. In the fullness of time the Second Law has its way, as organized matter is degraded back to water and carbon dioxide (CO_2) while the energy of light dribbles away as heat. The Second Law is the barb that points the arrow of time. But in the meantime we have the spectacle of a frantic society whose denizens jostle and compete for mates, water, space to breed, and places to hide, but especially for access to the sun's bounty.

The idea that ultimately energy flow is what holds together life's grand tapestry does not in any way conflict with evolution by heredity, variation, and natural selection. On the contrary, one view complements the other. As Jeffrey Wicken put it in his dense but seminal book,[9]

> The biosphere is an energy transformer, concentrating thermodynamic potential through the trapping of radiant energy and gradually degrading it to heat taken up by the sink of space. In so doing it necessarily provides niches in which [organisms] can emerge and evolve. [Organisms] are servants of the Second Law, means by which it expresses itself in the dissipative flow of nature. Conversely, the Second Law promotes the evolution of those pathways that can most effectively participate in, and support, the structuring flow of thermodynamic information in the biosphere.

Energy flow does not mandate the birds and the bees, nor does it explain their operations; what it does is to open up space for them to exist, and thus supplies a framework that supports the big picture.

Many Roads to Multicellularity

Every living thing large enough to see with the naked eye, with very few exceptions, is made up of eukaryotic cells, commonly in enormous numbers. The human body contains about 30 trillion human cells, plus another 30 to 40 trillion bacteria; the brain alone holds some 100 billion cells, about as many as the stars in our galaxy. Multicellular organisms make up as much as three-quarters of the world's biomass,[10] most of it in the form of land plants (bacteria come next with about 15%). Judging by quantity as well as quality, multicellulars represent a huge advance along the gyre of complexity. How that came about was until recently quite unknown, but thanks to molecular technology the fog has begun to lift.

The crucial discovery is that multicellularity is not a singular dramatic innovation (as the eukaryotic cell itself is thought to be), but happened repeatedly.[11] First forays in that direction may have been taken early on: some of the most ancient strata contain coiled filaments several centimeters in length that may be the remains of multicellular prokaryotes. Filamentous cyanobacteria made up of hundreds of cells, with some degree of specialization, date back more than 2 billion years. But multicellularity as a way of life hit its stride later, following the advent of the eukaryotes. The descendants of LECA multiplied prodigiously and diversified into an abundance of single-celled protists, presently classified into a small number of "supergroups" (Figure 8.2). Multicellular organisms are found in every one, displaying various sizes and degrees of organization, with three stems having proved particularly prolific: plants, fungi, and animals.

What makes multicellularity such a successful strategy? The benefits are conspicuous enough. Sheer size is one, which can be a deterrent to predators. It is probably more important that multicellular organisms are better at performing several specialized functions at the same time. Plant leaves, for instance, harvest light for energy and extract CO_2 from the atmosphere, even while the roots take up minerals from the soil. The division of labor is established during development, by a program of spatially regulated cell divisions designed to serve the interests of the organism as a whole. But the benefits come with a problem: cheaters. In a multicellular organism the component cells must surrender selfish reproduction for the good of the collective. Any cell that breaks the communal constraints undermines the whole; in humans we know this as cancer. Every multicellular organism represents a balance between benefits that accrue to the individual cells versus those that go to the community. In consequence, organisms differ in their degree of "multicellularity," with some ranking much higher on the scale than others. Animals and flowering plants stand at the high end, kelp with a minimum of specialized tissues come lower. Lichens come lower still, for the alga and fungus whose association creates the lichen can also go their separate ways.

In principle, multicellularity can be attained in two ways. One is aggregation of free-living cells, which is uncommon (the slugs of slime molds are one example). The more common way is for a multicellular organism to arise by regulated growth and division from a single founding cell, itself usually the product of sexual union. The most conspicuous benefit of a single-cell bottleneck at the start of an individual life is the elimination of cheaters: Cheaters can still jeopardize the life of the individual in which they arise, but will not normally be carried over into the next generation. This still leaves the hard work to be done: to establish a spatially extended network of communications by which cells at one location can inform, instruct, or constrain those at another. That is accomplished during the development of the fertilized egg into an embryo, and eventually into an adult. In Scott Gilbert's phrase, you can only get from genotype to phenotype via development.

Plants, fungi, and animals illustrate three very different modes of being successfully multicellular. Thanks to photosynthesis, plants are the primary producers of organic matter. They are sessile, their cells are enclosed in rigid walls and immobile, and they develop by the differential growth and division of those cells. Fungi, likewise, are protected by rigid walls and never mobile. Their characteristic form is the hypha, a slender filament made up of interconnected cells that extends exclusively at its tip. Fungi make their living as scavengers, relying on the secretion of digestive enzymes to break down organic matter into small fragments that can be absorbed; they are "osmotrophs." By contrast, animals and their component cells are typically mobile, with flexible frameworks and surfaces. They depend on the consumption of other living things: animals, and many protists also, take up organic matter in the form of solid particles and break it down internally. Unlike plants and fungi, animals eat. Their development, unlike that of plants and fungi, also entails cell movements. The very existence of such different modes of being multicellular suggested long ago that this transition, however momentous, came about more than once; and this inference has been amply confirmed.[12]

The search for the roots of multicellularity in our own animal lineage came to focus a century ago on an obscure group of flagellated protozoa called choanoflagellates. The reason is that their form is all but identical to those of the collar of feeding cells that line the internal channels of sponges, which are thought to represent the most ancient of all animals. The recognition of choanoflagellates as the closest living unicellular relatives of sponges has been strongly reinforced in recent years by molecular and genomic data. Briefly, it used to be thought that the characteristic features of animals all evolved after they became multicellular, but that is evidently not quite so. Choanoflagellates possess a suite of proteins and genes clearly related to those that participate in animal development, including proteins involved in cell adhesion, intercellular communications, and the regulation of gene expression. These, it seems, were co-opted and repurposed during the emergence of the animal lineage. The findings should not be construed to suggest that animals evolved from choanoflagellates; rather, these two lineages share a common ancestry (Figure 8.2). Multicellularity in the plant lineage evolved quite separately from that of animals. The closest unicellular relatives of land plants are green algae (e.g., *Chlamydomonas*), with which they share many genes. The multicellular red and brown algae likewise have identifiable unicellular cousins, as do the fungi (chytrids). However momentous its consequences, the transition from unicellular to multicellular came often and quite easily.

Now comes the hard part. There remains a huge gap between choanoflagellates and say, roundworms; and it remains hard to fathom how metazoans (multicellulars of the animal lineage) fit to conquer the world came into existence. The first ancestral metazoan seems to have been, not only multicellular but made of differentiated cells arrayed in sheets; it had a body plan and regulated development that included some kind of gastrulation. A hypothesis of metazoan evolution, formulated by Ernst Haeckel in 1874 and still widely accepted, envisages the ancestral metazoan as a hollow ball of flagellated cells that turned inside-out, as colonies of *Volvox* still do today. It would have lived by catching bacteria with its fringe of collar cells, reproduced by means of sperm and eggs, and had some capacity for morphogenesis, possibly with the help of stem cells and selective cell death. This will have been a creature as different from a unicellular protist as LECA was from its prokaryotic progenitors. Its genesis called for much creative novelty, focusing not so much on novel genes as on the reformulation of regulatory mechanisms, and that is where we go next.

Sources of Novelty

The history of life is speckled with novelties that vary in kind and magnitude: adaptations such as resistance to some disease or variation in color pattern; innovations including flowers, seeds, scales, and feathers; and rare but radical transformations, including the eukaryotic mode of cellular organization. Museums of natural history make a point of highlighting conspicuous novelties such as legs, wings, and eyes with lenses and retinas. The origin of novelty has been one of biology's major challenges ever since Darwin, and it continues to stir debate.

The standard conception of novelty was formulated seventy years ago as part of the modern synthesis, which melded Darwin's insights from natural history with the rising science of genetics (Chapter 5). It envisaged a static genome composed of discrete heritable genes that undergo occasional variation by mutations and other accidents. Heritable variations are culled by natural selection, which eliminates harmful ones and preserves the few that promote adaptation to changing circumstances. New functions begin with accidental changes in a single gene, are augmented by subsequent mutations that generate useful features gradually by small random steps, and are shaped throughout by positive selection for improved function. Macroevolution, the origin of phyla and kingdom, occurs by the same processes as microevolution. Extinction is the fate of lineages that fail to keep up with the competition, or that fall victim to some catastrophe. Skeptics have questioned this scenario from the beginning, arguing that there never was time enough to bring forth the profusion of novelty by such pedestrian means, and in any event random mutations are more likely to degrade organization than to create it. The growing knowledge of genes, genomes, and physiology has confirmed the classical view, while also reinforcing the sense that the production of novelty quite often follows from leaps rather than creeps.

One of the virtues of Bacteria as experimental subjects is that they multiply quickly enough to let microbiologists catch events on the wing. It is possible to spot rare mutations as they arise, and to trace both their origins and their consequences in molecular detail. Such studies leave no doubt that the traditional account correctly describes one way by which new adaptations come about. A remarkable example is the manner in which, on rare occasions, a population of *Escherichia coli* can acquire the capacity to ferment citrate. This requires not just a single very uncommon mutation but at least two more preparatory ones.[13] But research on microorganisms has also demonstrated that there are many roads to innovation, some of them distinctly unconventional.

Genomes, prokaryotic ones in particular, turn out to be much less static than was formerly imagined. On the contrary, on the timescale of hundreds or thousands of generations they are subject to remodeling by viruses that jump into and out of position, by a drizzle of foreign genetic material picked up from the environment, and even by cellular enzymes that reconfigure the genetic instructions.[14] Genomes turned out to be more dynamic than expected, more malleable and interactive, and therefore more evolvable. We also understand that not all new features are adaptations that promote their bearer's survival and reproduction. There is a spontaneous tendency for genomes to become larger and more complex over time, not because this is called for by natural selection but because of inevitable errors in replication that lead to elaboration and duplication. Such events, unless overtly harmful, can become fixed and perpetuated even in the absence of positive selection, and may later serve as foundations for some useful feature. Some of the baroque complexity of living things, particularly in eukaryotes, may have originated in this manner.

Cells are dynamic systems, shaped by networks of genes whose products organize themselves into higher-order structures (Chapter 3). This becomes prominent in higher organisms: evolution of a new function or structure requires the generation of a modified (or brand-new) mesh of signals, receptors, and effectors. Altogether new genes are not common,[15] but changes in the pattern of regulation are involved in all innovations, extending over hundreds of protein-coding genes (Chapter 10). Among the most common are mutations at regulatory loci that may specify either a protein or one of the newly discovered kinds of regulatory RNA, and affect the timing, rate, or extent of some process of developmental significance.

The complexity of these interactions peaked with the discovery of "epigenetic" effects, heritable alterations in exposed parts of DNA that affect the expression of a genetic message rather than its content.[16] These are known to play a large role in the development of embryos: a liver cell has the same genome as a skin cell, but epigenetic markers suppress the expression of one set of genes while activating another. To what extent epigenetic changes contribute to evolution is very much under debate. They are easily reversed by cellular enzymes and therefore lack the durability of a change in DNA sequence, but many scientists suspect that epigenetic changes can be assimilated into the genome proper. If so, here would be a way in which physiological responses to stress can be made permanent, and open the way to the inheritance of acquired characteristics. Classical evolutionists consider that a major heresy that would pose a challenge to the entire doctrine, but the field seems to be taking it in its stride.

The most spectacular innovations are associated with symbioses, which bring together unrelated organisms into associations of varying degrees of intimacy and commitment.

Lichens, symbioses of a fungus and an alga, flourish on bare rock, which neither partner can do on its own. Intracellular green algae living in the tissues of corals enables the latter, animals though they are, to profit from photosynthesis. But this is a marriage of convenience: a small rise in the ambient temperature will dissolve the partnership, causing the "bleaching" that is killing corals the world over. Permanent, indissoluble symbiosis is rare, but it underpins some of the most dramatic innovations in the expansion of life. As we saw in earlier chapters, the evolution of the eukaryotic cell entailed two of them: first the acquisition of Bacterial symbionts that gave rise to mitochondria, and then a second and more restricted symbiosis that produced plastids.

With so many ways to generate novel features, is it possible that organisms can purposefully rewrite their own genomes in response to particular challenges or opportunities? That, if true, would really shake the foundations of evolutionary thought! Claims to this effect have been put forward more than once,[17] but none have held up. There seems to be no such thing as "directed mutations," targeted genetic changes that respond in a specific manner to stimuli from the environment. All mutations appear to be "random," in the sense that they are chance events that happen when they happen, regardless of the needs of the organism. By the same token, and despite the current enthusiasm for epigenetic inheritance, changes that are not grounded in nucleotide sequences are apt to be lost within a generation or two. Permanent innovations, it seems, must be inscribed in the genetic makeup by one means or another, and ratified by natural selection. What we marvel at on the level of wings or seeds is always the outcome of protracted evolution in the classical manner.

All right then, is selection for improved function the sole, or even the chief, driving force behind the appearance of novel structures and functions? That has been the prevailing view for almost a century, and no one claims that intricate and purposeful organs can arise in the absence of natural selection to lend direction to the chaos of change. But recent years have brought a growing appreciation of neutral or nonadaptive evolution, which underlies the broad trend toward increasing complexity and may even come to mimic the effects of natural selection.[18] All claims to purpose-driven evolution must be judged against this background of directionless change.

This is important, for the status of evolution as the central organizing principle of biology hinges on it. Over the past half-century traditional views have loosened, and that has fueled calls for expansion of the modern synthesis, even its replacement. As I see it, expansion is happening without fanfare all across the field and that is all to the good, yet Darwin's place on his pedestal remains secure. A theory that accommodates lateral gene transfer, jumping genes, and the occasional inheritance of acquired characteristics is not what Darwin would have imagined. But so long as we construe the history of life as a story about heredity, variation, and natural selection, we see the world through Darwin's eyes.

The Pattern of Proliferation

Forty years ago a pair of distinguished paleontologists tossed a stone into the placid pond of evolutionary thought, starting ripples that continue to spread. J. S. Gould and N. Eldredge[19] asked whether the fossil record of animals confirms the conventional expectation that evolution is slow and gradual, the outcome of numerous small changes over a long period of time, and concluded that it did not. There is no indication that new forms arise by the gradual transformation of their predecessors. Instead, change is episodic or "punctuated," and linked to the appearance of new species. Species show up suddenly, endure for some millions of years and then vanish as abruptly as they appeared. Subsequent research has confirmed and extended the conclusion, particularly into the microbial world. Not species alone but major groupings as well appear in bursts that retain their pattern of order over long periods of time, and are best regarded as the exploration of variations on a successful new mode of organization (Figure 8.1).

Initially, some scientists argued that the punctuated pattern of life's proliferation contradicts the principle of evolution by variation and natural selection, but that is not so. In the fossil record, a few inches of rock may represent many thousands of years, plenty of time for the emergence of new species in what looks like the blink of a (geologist's) eye. It has been estimated that it takes as much as 50,000 to 100,000 years for a new species to emerge. What Gould and Eldridge had uncovered is not a flaw in Darwin's theory but a pattern of large-scale order that only becomes apparent on the geological scale of time.

Why is evolution linked to speciation? An episodic pattern of expansion is what would be anticipated on the premise that what promotes expansion is the discovery of new ways to produce useful substances, harvest energy, organize and reproduce structures—in a phrase, new ways of being in the world. Species appear, strut their hour on the stage and then yield to others that are by some measure "superior." Species come and go; the technology that underpins their lives is likely to endure. Most inventions are restricted to some particular lineage (the spiracles that allow insects to breathe, for example). Some appear repeatedly: eyes were invented more than a dozen times. And some are of such magnitude that they alter the very course of biological history, like photosynthesis. Figure 8.1 collects some of the transitions, or turning points, in evolution.[20] They include, first of all, the appearance of self-reproducing molecular systems, the organization of cells, and the invention of the major metabolic pathways. A prominent feature is the rise of larger and more inclusive individuals, such that constituents that once reproduced independently could now do so only as part of some larger entity; the rise of the eukaryotes is a prime example. Subsequent transitions are more familiar but narrower in scope, with the possible exception of the appearance of language in a single lineage of primates, whose consequences for the entire biosphere are even now beginning to play out.

Of all the turning points of biological history, none has drawn more puzzled consideration than the "Cambrian Explosion." Briefly, abundant animal fossils appear "suddenly" at the beginning of the Cambrian era, 542 million years ago, representing all the phyla (body plans) known today. Darwin was deeply troubled by the fact that Cambrian strata teem with fossils, while those just a little earlier are barren. Did that not call into question his entire theory, predicated on gradual change over time? Contemporary investigations of Precambrian strata inaccessible to Darwin, notably in Australia, Siberia, and Mongolia, have laid to rest those doubts but raised others.[21] It turns out that the many layers of Precambrian rocks are not really devoid of life. They record a sparse but widespread fauna of tube worms, snails, sponges, even pioneering predators equipped with teeth that could pierce the defensive shells evolved by their prey. Arms races are not a modern plague! And yet, in another sense the Cambrian Explosion is real enough: modern body plans, exemplified by arthropods (trilobites, particularly), mollusks, and vertebrates, all appeared during a narrow window of time at the start of the Cambrian. Just what happened then remains one of biology's many riddles.

Most puzzling of all the Precambrian creatures are the Ediacarans, originally found on a station by that name in Australia but now known to have lived around the world some 600 million years ago. Large, flat, leaf-like objects lacking limbs, mouth and anus—what could these have been, multicellular animals of a design that has been lost? If so, what and how did they eat, and how did they dispose of waste? What is their relationship to the more conventional animals of Cambrian times? Controversy continues, but it is clear that the Cambrian Explosion was more protracted and more complicated than the term suggests.

Does Evolution Have a Direction?

On the face of it, the answer is self-evident: over the span of nearly 4 billion years, life has burgeoned both in quantity and in quality. Once there were only bacteria, and now there are elephants, forests, and all the productions of humans from science and symphonies to mountains of plastic trash. And yet, when biologists examine lineages of either plants or animals they find no evidence of directional change over time. Nor is there anything in the process of evolution by heredity, variation, and selection that predicts long-term progression—yet there it is. What is going on?

Whatever it is, it occurs episodically at the nodes where new species arise, often in the aftermath of an extinction. It is definitely not a pointed advance toward a single peak (e.g., mankind, as in the medieval image of The Great Chain of Being), but one that has produced a great bush of lineages displaying different sorts of excellence of different kinds and of varying degrees. There are excellent worms, no less than excellent flowering shrubs. To the question, just what is it that has increased so spectacularly over time, the first answer that comes to mind is "complexity": more and larger organelles, more kinds of cells, more connections and modes of interaction; and surely this is true. We have no objective way of measuring complexity (or even to define it in the biological context), but no one will deny that an elephant is more complex than a cell of *E. coli*. Moreover, we are gaining insight into how and why the number of parts and connections tends to increase over time, and it is not necessarily a matter of selection and adaptation. The larger and more intricate ribosomes of eukaryotes have no obvious selective advantage over the more streamlined prokaryotic ones. Instead, it seems that genes are subject to errors during replication that tend to make them longer, or even duplicate; and in small populations such genetic changes can be fixed by "drift," even if they confer no advantage and may even be slightly deleterious. It seems that a tendency operates in biological systems that is reminiscent of Parkinson's Law: Complexity increases to fill the space of accessible possibilities.

So far so good, but a spontaneous increase in the number of parts and interactions cannot be all the story. "Complexity" serves as proxy for something less tangible but more profound: the sense that over time there has been a dramatic increase in functional, purposeful organization, and that this is what the evolutionary drama is all about. It is hardly disputable that over time there has been a global trend toward increasingly intensive energy metabolism, more hierarchical organization, and enhanced physiological capacity, and also toward complexity as defined earlier. Moreover, though nothing in the theory mandates such a trend, most biologists are persuaded (as was Darwin, who was well aware of the issue) that a global "progressive" trend is ultimately the outcome of variation, natural selection, and adaptation. "Neutral evolution," change without selection, is a real factor as well, but it seems inconceivable (at least to me) that organs of perfection from cilia to eyes could have come into existence without selection to confer direction on the cacophony of random change. Natural selection appears to be both necessary and sufficient: there is no evidence for, nor any need to invoke, something more directional to bias the outcome at critical junctions, let alone God's finger on the scale.

We clearly need to find words that highlight the quality that evolution seeks to maximize. Two terms from the contemporary literature give me some sense of what evolutionary progress means.[22] One is Stuart Kauffman's precept of "agency," an organism's capacity to act on its own behalf. The other is Bernd Rosslenbroich's usage of "autonomy" to designate the manner in which the direct influence of the environment comes to be progressively reduced, while the flexibility and stability of factors intrinsic to the organism are enhanced. A good example is the ability to regulate body temperature, prominent in mammals and birds but absent in reptiles (dinosaurs may have been an exception), in plants, and all unicellular organisms.

Well then, do we now understand how and why life has expanded so prodigiously, and does Darwinian evolution live up to its billing as the unifying principle that alone makes sense of the living world? I think so, but need hardly remind readers that this 150-year-old idea continues to generate conflict, particularly in America, where half the public rejects the very principle. For those impressed by evidence and reason, the case is irrefutable, and what remains to be discovered comes under the heading of puzzles, not mysteries. The major exceptions are the origin of life and the workings of mind, and these ought to be sufficient for lovers of the profoundly baffling to chew on.

9

The Tangled Bank

It is interesting to contemplate a tangled bank, clothed with many plants of many kinds, with birds singing in the bushes, with various insects flitting about, and with worms crawling through the damp earth, and to reflect that these elaborately constructed forms, so different from each other, and dependent upon each other in so complex a manner, have all been produced by laws acting around us. These laws, taken in the largest sense, being Growth with reproduction, Inheritance . . . , Variability . . . , a Ratio of Increase so high as to lead to a Struggle for Life, and as a consequence to Natural Selection, entailing Divergence of Character and Extinction of less improved forms. Thus, from the war of nature, from famine and death, . . . the production of the higher animals directly follows.

—Charles Darwin, *The Origin of Species*[1]

The biological conversation is freighted with detail and arcane controversies, but its underlying questions are commonly plain and straightforward, the kind a bright teenager might ask. Why are there so many kinds of animals or plants? Why do they cluster into discrete groups rather than spreading into a continuous smear? Why do organisms living in one place differ from those in another? The classical answers are mostly there in the final paragraph of Darwin's *Origin of Species* (1859), probably the best-loved of all his writings.

The continuing effort to spell out just how the tangled bank grows out of the principles of heredity, variation, competition, and natural selection feeds into the expansion of the evolutionary landscape. As a result we have come to appreciate that if competition is one facet of life, cooperation is another. These two are the Yin and the Yang of evolution, and to grasp the whole we must recognize their interdependence.

Why Are There So Many Kinds of Organisms?

The diversity of living things is their most conspicuous feature, it all but hits one in the face. The variety of mammals is fairly modest. Zoologists estimate some 5.000 species, of which 1,500 are rodents; there are but 4 or 5 (sub)species of giraffes, and a single surviving species of humans. By contrast, plants, insects, and protists are prolific. There may be over 300,000 species of land plants, 5 to 10 million of insects (the count of beetles alone exceeds 300,000), and another quarter million unicellular eukaryotes. No one knows how many kinds of prokaryotes exist, but the number is surely in the millions, most of which cannot be grown in culture.[2] The most obscure are the denizens of the "deep hot biosphere" (Thomas Gold), Bacteria and Archaea that inhabit porous rocks deep underground and sediments beneath the ocean floor. They tend to be tiny, exceedingly slow growing (generation times in the hundreds of years!), and subsist on geochemical energy or the drizzle of organic debris that filters down from the surface. Populations are sparse, but, thanks to the huge volume of their habitat, these cryptic creatures may make up a third of the world's biomass.[3] Yet even these exotic forms appear to be no more than variations on the universal biochemical themes, and all nestle within the three recognized domains of Bacteria, Archaea, and Eukarya. Despite assiduous search, no fourth domain has been found.

Biologists are unanimous that both diversity and its grainy texture result from the interplay of heredity, variation, and natural selection, but argue fiercely over the details.

We need not get distracted by the bickering over just what makes a "species." Zoologists consider that organisms that can interbreed to produce fertile offspring belong to the same species; botanists and microbiologists necessarily employ different criteria. For present purposes we can think of a species as a cluster of organisms (populations, really) united by common descent and by similarities of form and function, and objectively distinguishable from other such clusters. Each cluster represents, in Ernst Mayr's phrase, "the product of a well-balanced, internally cohesive genotype"[4] that remains stable over a long span of time, sometimes running into the millions of years.

What makes a harmonious and persistent pattern of traits is adaptation to some particular niche, a way of making a living. The world is patchy, made up of innumerable locales that differ in physical and chemical characteristics (temperature, acidity, salinity, and so on), each of which favors some combination of genes over others. Over time, the proliferation of living things created an ever-growing range of niches around organisms, on their surface, and even within other organisms themselves. Microbes in particular commonly populate microniches, specialized habitats that may be physical, nutritional, even behavioral. Mutations and other changes to the genotype are winnowed by natural selection, which improves adaptation and sharpens the cluster. Sexual reproduction reinforces the barriers between one cluster and another. We can say, then, that the granular texture of the living world is to a large extent a product of the way it came into existence.

Should we then conclude that every trait and feature enhances survival and brings some selective benefit? Both common sense and the evidence suggest that this is not the case. More likely, every species displays a blend of features that have some utility with others that do not. Morphology is a happy hunting ground for both sorts, especially as it is often very difficult to assess the benefits of a particular shape. The most conspicuous features are surely products of selection for better performance: the flagella that make many bacterial cells motile, the leaves that harvest light as a source of energy, echolocation in bats, or the torpedo shapes of sharks and tuna. But most biologists today reject the claim that every feature is there for a purpose. There are always alternative ways of doing things, whose advantages and costs are not obvious. In eukaryotes particularly, the complexity of cell structure tends to increase over time for reasons unrelated to economic benefit. The profusion of forms suggests nothing so much as sheer playfulness, an exploration of the endless possibilities of "morphospace," many of which may be neutral from the standpoint of natural selection.

The classical view of biological diversity sketched here takes all that comes between genes and organisms as a black box, and sets it aside. The presumption is that, although we don't know what all is in the box, for the purpose of parsing evolution that does not much matter because only what's bred in the genes comes out in the flesh. Similarities between organisms are due to common descent, and the closer the family ties the more traits will be shared.

Diversity is the result of alterations to the genome, small and large, which ultimately account for the characteristics of organisms, including novel devices, adaptation, and the proliferation of forms. Evolution has no foresight, no plan, and no purpose, therefore any form or function goes as long as it is endorsed by natural selection. There clearly is much truth in this way of looking on life but not all the truth, and the worm in the apple lurks in the gap between genes and organisms. We saw in previous chapters that genes specify the chemistry of organisms (albeit sometimes quite indirectly), but not the higher levels of order, including form. Those arise as expressions of the dynamics of self-organizing systems, and it is this level that gives rise to the integrity, the "wholeness" that strikes every observer of living things. We should therefore expect the course of evolution to display the effects of two kinds of constraint: those that stem from the conservatism of the genome, and others rooted in the physical workings of each particular system.

This is a distinctly unfashionable point of view that implies yet another enlargement of the Darwinian envelope; is there any evidence to support the assertion that system dynamics guide and constrain the forms that can emerge in the course of evolution? Yes, there is some, but it is not nearly as compelling as that which underpins the link to the genes. Most of the evidence is indirect, and some goes back to an earlier stage of biological thinking. For example, D'Arcy Thompson noted a hundred years ago that the shapes of diverse fishes can be derived, one from another, by plotting the shapes on a grid and then distorting the grid. It is hard to rationalize observations of this kind in terms of individual genes, but they make good sense on the premise that an organism is a flexible system that responds as a unit to pressures both from within and from the environment. Contemporary biologists model the shapes of cells and organisms on a computer screen, and try to derive them from the manner in which these forms are actually generated. In the hands of the late Brian Goodwin,[5] that has led to the formulation of morphogenesis as the expression of a physical "field," a region of space over which forces act in a coherent manner. This point of view as been highly productive at the physiological level, and it suggests that on the evolutionary timescale organisms illustrate variations on a large (but not infinite) range of "generic forms." I believe that this approach holds much promise, but it stands well outside the mainstream. The quest for a rational basis to forms and their evolution still has a long way to go.

If we accept Mayr's thesis that each species is the expression of a balanced and cohesive genotype, what makes them diverge and differentiate?

Generally speaking, divergence begins with the rise of some sort of barrier that isolates part of a homogeneous population from the majority. In the absence of barriers members of the whole population interbreed, and the resulting flow of genes prevents the emergence of stable subpopulations. Barriers may be physiological (a novel item of diet, perhaps, or a local disease), or behavioral, but they are most commonly geographic. A handful of finches that managed somehow to reach the Galapagos Islands, far off the coast of Ecuador, founded a population isolated from the finches of the mainland. Over time these differentiated into local populations characteristic of the several islands, each with its own favorite diet, shape of beak, color pattern, and typical behavior. Like politics, species formation is always local.

Diversity begins small and local, but it grows with the years. Given time enough, selective pressures and the appearance of novel adaptations, differences may deepen and eventually come to reach beyond the level of a species. So far as we know, the profound differences that now separate birds from their ancestors among the dinosaurs began, more than a hundred million years ago, with small innovations that separated one population from the rest and gave rise to a new species. In this view macroevolution, the genesis of the major divisions of life, is not different in essence from the microevolution that produces new species—it has just been at it longer. As Darwin put it, what is now the stout limb of a great tree was once but a twig on a slender sapling. As usual there is much truth in this but not all the truth, and the glaring exceptions turn on symbiosis.

The Cooperative Side of Life

The Darwinian outlook highlights individual organisms, in competition with other individuals of the same or other species, as the targets of natural selection. Cooperative relationships played little role in the formulation of the modern synthesis, even though by the middle of the 20th century the existence of various kinds of symbiosis ("living together") was well recognized. Symbioses were taken to be biological curiosities, not a fundamental aspect of how the world works. The broadening of evolutionary attitudes has opened a wider space: just as human history was shaped as much by trade as by warfare, so cooperation and competition are complementary aspects of the ascent of life.

Associations differ widely in kind and intimacy, ranging from mere co-habitation to mutual services and total interdependence. Corals rely on one or another kind of green algae to supply more than half their energy needs. When the water warms the partnership breaks down, the algae are expelled and the "bleached" coral dies. Legumes are well known for their ability to fix atmospheric nitrogen and replenish nitrogen-depleted soil. They owe that skill to symbiotic bacteria housed in specialized nodules in the roots of the plant. Mycorrhizal fungi tucked in among the roots of forest trees supply the latter with minerals in exchange for products of photosynthesis. Perhaps the most spectacular instances of symbiosis are the lichens, associations of a fungus with either a green alga or a cyanobacterium. The partners can live independently, but only the consortium thrives on the bark of trees, even on bare rock; and only the consortium takes on the characteristic forms of lichens. Humans, also, live in symbiosis with the many kinds of bacteria that flourish on and within us. Embryos are sterile but begin to pick up bacteria even as they are born, and in the absence of the mixed flora that populates our guts the immune system fails to develop properly.

What ultimately brought symbiosis from the margins into the heart of evolution was the discovery that the eukaryotic cell itself is the product of a most intimate kind of symbiosis, intracellular or endosymbiosis. Protists commonly harbor endosymbiotic bacteria, and serve as models for how the most far-reaching transformation in cellular organization may have begun. As outlined in Chapter 6, mitochondria, the powerhouses of eukaryotic cells, are descendants of alpha-proteobacteria that long ago took up residence in a host cell of Archaeal affinities, and played a major role in shaping the eukaryotic order. A subsequent and independent act of endosymbiosis brought a cyanobacterium into an early branch of the eukaryotic tree; this evolved into plastids and initiated the spectacular proliferation of photosynthetic eukaryotes. Later on, a series of secondary and tertiary endosymbiosis spread photosynthesis all across the eukaryotic universe. There is nothing marginal about symbiosis: were it not for endosymbiosis, the living world as we know it would not exist. Some would go even further, especially the late Lynn Margulis, who insisted that symbiosis, rather than competition, is the chief driver of the evolutionary plot. Few biologists subscribe to that radical view, but the recognition that symbiosis provides an alternative and complement to neo-Darwinism alters the way we think about the very nature of living things.

What makes an organism? Considering that life is dispensed in units that are universally referred to as organisms, it comes as a surprise to learn that biology has no formal definition of the term. Still, there is fairly general agreement about what we mean: "A functional and cohesive whole made up of interdependent and interconnected parts."[6] What distinguishes organisms from other physical systems is that they meet the criteria of being alive (Chapter 1), and most pointedly that the parts work together for the benefit of the integrated whole. Evolution has "designed" them for an apparent purpose, to persist, multiply, and inherit the earth. This conception is flexible enough to encompass both the lichen and its constituent fungus and alga, and allows for multiple levels and degrees of "organismality." More troublesome is the commonplace notion of an "individual." The classical understanding, that a single individual correspond to a single genome, fails when an organism is made up of more than one genome. For an extreme case, consider the Australian termite *Mastotermes darwiniensis*, which digests wood and builds elaborate underground nests. Both these remarkable skills are totally dependent on the presence in the termite's gut of a peculiar protozoan, *Mixotricha paradoxa*, itself a consortium of five genomes (a nucleus plus four endosymbiotic bacteria, but no mitochondria). Where in this menagerie do we find the individual termite? What is the unit on which selection acts?

Scott Gilbert and his colleagues argue persuasively that the focus on individuals defined by the possession of a single genome is mistaken from the start. For most if not all organisms larger than bacteria, a more natural unit is the "holobiont," the multicellular ensemble of eukaryotic cells plus its complement of persistent symbionts. This is the "central unit of anatomy, development, physiology and immunology,"[7] that functions as a cohesive unit in interactions with its environment, and is normally the target of natural selection. It is the holobiont alone, not any one of its constituent organisms, that possesses the complete genome. Adoption of the symbiotic perspective calls for yet another enlargement of the evolutionary landscape. If the lichen, as well as the fungus and the alga, is a target of natural selection the door falls open to "group selection," an idea that has long been repugnant to neo-Darwinists (except in the context of social groups). In a profound sense, higher organisms, including humans, "have never been individuals."[8]

All Together Now

Nature recycles; all the productions and fabric of living things are ultimately broken down, decomposed, and returned to the dust from which they emerged. Nature's frugal economy operates by way of the tangled web of organisms, each of which participates in its own particular and local manner. The waste products, and the bodies, of one kind are commonly another kind's daily food, with fungi and bacteria as the recyclers-in-chief. At the same time, living things live in larger assemblages, ecosystems, that create niches for one another; and they take an active part in shaping their environment. What has expanded so prodigiously over time is not some particular class of organisms but the web of life, that network of organized biochemistry that has spawned endless diversity from a base of chemical unity. That, in the end, is the reason for Walter de la Mer's "odd fact / as odd as it can be / that everything Miss T. eats / turns into Miss T.": Miss T. is made of the same substance as her biscuits and her tea.

We humans, eukaryotes and higher organisms, take pride in the superior capacities of our kind; but when it comes to nature's economy her chief instruments are microbes, and so it has been from the beginning. Earth is an open system, bathed in a torrent of energy from the sun, and it is microorganisms which harvest that energy and feed it into the universal biochemical web. Even the contribution that geochemical energy makes to the biological budget is funneled by microbes, such as the Archaea that generate methane and the Bacteria that consume it. The mechanisms that harvest energy and transduce it into usable forms are all prokaryotic, and have been strongly conserved for at least 3 billion years. Eukaryotes rely on mitochondria and plastids, descendants of enslaved Bacteria, and have invented no novel mechanisms. With regard to matter, unlike energy, earth is very nearly a closed system (meteorites excepted), and so the living world must continually recycle its elementary constituents: carbon, nitrogen, oxygen, sulfur, and phosphorus. Weathering releases some of these from rocks, but the largest contributions come to us courtesy of an array of microbes.

For an example, take nitrogen. The major constituent of the atmosphere, nitrogen is also a component of proteins, nucleic acids, and many other biomolecules, but it is not easily come by. Aside from what nitrogen is supplied by organisms living and dead, the only portal into the biochemical web runs through ammonia, a highly reduced form of nitrogen which is wholly of microbial provenance. Some Bacteria and Archaea contain the enzyme nitrogenase that catalyzes the reduction of nitrogen to ammonia, a very costly process that requires no fewer than twelve molecules of ATP for every atom of nitrogen, and gives its possessors an edge in nitrogen-poor environments. Other prokaryotes make a living from reactions that oxidize ammonia and other nitrogenous substances, and return nitrogen gas to the atmosphere. As Paul Falkowski says, it is microbes that make the world habitable for all higher creatures.[9]

The persistence of metabolic cycles that conserve nitrogen, sulfur, and phosphorus pose a riddle for evolutionary theorists. There is geological evidence that nutrient cycles, in principle similar to those of the present, but probably mediated by different prokaryotic players, go back to the very dawn of cellular life. What has allowed them to endure for billions of years? The glib answer would be natural selection, but evolution acts on self-reproducing and variable individuals; can large-scale networks of reactions also be subject to selection, and if so, how? That is one aspect of a still larger question: How has Earth remained a habitable planet for all of geological time despite all the changes it has undergone? This has been quite a lively issue, to which we turn shortly.

But first, a nod to an inconvenient and ominous truth. Nature recycles, and in olden days that was true of human societies as well. Those hunters, gatherers, and farmers that make up the great bulk of human history were solidly tied into the economy of nature. Only with the industrial revolution, little more than two centuries ago, did we begin in a serious way to diverge from past norms. We glory in our wealth, technological prowess, knowledge, and liberties, but these come with wasteful consumption, pollution, a tide of plastic trash, and the proliferation of humans at the expense of all the rest of the natural world. Half of all the nitrogen that now enters the circulation comes from the industrial reduction of nitrogen gas, upsetting the economy of lakes and waterways. Climate change is but the latest indicator that we are on a course that is not sustainable; but that is a message that we refuse to hear, let alone heed.

The Self-Regulating Earth

Strange, is it not, that despite all the climatic and geological upheavals of the past four billion years, Earth has avoided the fate of our neighbors Venus and Mars and remained continuously hospitable to life? The increase in the young sun's energy output, meteorite impacts, massive lava flows, global ice ages, and the oxygenation of the atmosphere—life took it all in stride and recovered quickly from every perturbation. Given the notorious fragility of living things, it is hard to rationalize the resilience of the biosphere. Indeed, many scientists hold that there is nothing to explain, it's all a matter of sheer luck that could run out without notice. There must be some truth in this assessment: if the giant meteorite that slammed into the earth 66 million years ago and put paid to the dinosaurs had been larger, it would have incinerated the entire planet. Even so, the steadfast durability of Earth's environment cries out for explanation beyond good luck alone.

In the 1960s this puzzle drew the attention of James Lovelock (b. 1919), a noted atmospheric chemist and inventor of instrumentation. Starting from the insight that Earth's atmosphere, containing methane as well as oxygen, is grossly out of chemical equilibrium, and that the continuous consumption and replenishment of these gases is due to the activities of living things, he came to conceive of our planet whole as a system of linked chemical reactions that together maintain it in a state favorable to life. Lovelock's ruminations, sharpened by collaboration with Lynn Margulis, crystallized in the concept of Earth as a self-organized web of coupled processes, both biological and geological, that sustains and stabilizes itself. Collectively, multiple unrelated processes mesh together so as to keep the composition of the atmosphere, the chemistry of the ocean, and global temperature within boundaries compatible with life. To highlight the central idea, Lovelock adopted the name of an earth-goddess of ancient Greece, Gaia, and thereby caught the public imagination. His evolving views were spelled out in technical articles and popular books, and triggered an academic tempest.[10]

Gaia came on the scene at a time of rising public awareness and concern over pollution and the destruction of natural habitats, all in the name of free enterprise and economic growth; and she suited the mood perfectly. "Earth is alive!" Rachel Carson's *Silent Spring* was published in 1962, the Wilderness Act passed Congress in 1964, and Earth Day was first celebrated in 1970. Gaia became an icon of the environmental movement, and remains so. But the reaction of the scientific community, much to the chagrin of Lovelock and Margulis, was reserved—even hostile. The conception of Earth as a self-regulating whole appealed to a minority attracted by holism and systems-thinking (myself included), but was altogether out of step with the pervasive worship of reductionism, soon to culminate in the rise of gene-centered molecular biology. Biologists were disturbed by the lack of positive evidence for homeostasis on the planetary scale, and repelled by the personification of an abstract hypothesis as a pagan goddess. Professional evolutionists were especially scathing, insisting that a singular, nonreproducing entity such as a self-regulating earth could never come into existence by an evolutionary process, since those require heredity, variation, and natural selection. The conventional attitude among biologists remains thoroughly skeptical.

Yet even in the groves of academe, the times they are a-changing. There is general agreement that Earth is not a living organism in the sense that bacteria or bumblebees are alive, but it is also now widely understood that complex systems cannot be grasped in terms of their molecular parts alone. Moreover, the insight that the planet represents a self-regulating and self-sustaining system has gained credibility as some of the strands in that web of interactions have been unraveled. To take one instance that has been much discussed, consider the role of planktonic algae in stabilizing the global climate. Warming promotes growth of the algae, which produce dimethyl sulfide (by degradation of a substance that serves as an osmotic stabilizer). Dimethyl sulfide is released into the atmosphere, there to be oxidized to sulfate. Sulfate contributes to the formation of aerosols, which are a major source of nuclei that promote the condensation of clouds. Clouds, in turn, reflect solar radiation back into space, which tends to cool the atmosphere and restore equilibrium.

On a still larger scale, life interacts with geological processes. Volcanoes spew vast quantities of CO_2 into the atmosphere, which tends to raise the global temperature via the greenhouse effect. Weathering of rocks ties up CO_2 in the form of carbonates, which are washed into the sea, there to be taken up by marine algae to construct their intricate shells. When the algae die, the shells settle onto the seafloor, to be compacted into limestone and buried—thus withdrawing CO_2 from circulation. But the cycle does not end there: plate tectonics draw those rocks into the depths, where they melt and supply the CO_2 that volcanoes discharge into the atmosphere. According to Gaia theory, the network of interdigitating cycles acts as a gigantic feedback loop that stabilizes the physiology of the planet: "atmospheric homeostasis by and for the biosphere." No doubt, many more links remain to be discovered. But it is already clear that the system's capacity to maintain the present state is not unlimited, and is now coming under increasing stress from human activities.

The question remains, how could a self-stabilizing global system come into existence by natural causes? The organisms that participate in it are themselves ruled by heredity, variation, and natural selection, but that cannot apply to Earth as a whole.[11] As Tim Lenton asks, how can planetary self-regulation emerge from selection at the level of organisms? The answer, it appears, may be that what evolution rewards is persistence itself. We have already encountered this idea in the context of the origin of life (Chapter 5), and here it is again on the planetary scale. W. Ford Doolittle, once a fierce critic of Gaia theory, rationalizes the emergence of a self-stabilizing network by emphasizing the benefits, not to the organisms that participate in the web, but to the web overall.[12] The activities of those organisms, each a product of Darwinian selection in the conventional manner, support the endurance of the web; and the longer it endures, the more likely it is to accumulate additional strands that promote its persistence. Should the web be disrupted by the extinction of one of the players, it is likely that members of some other group will fill the vacant niche and maintain the global order.

All this sounds sensible enough, at least to me. I take Earth to operate as a self-sustaining and self-regulating system, made up of both biotic and abiotic strands, rather as a human city is. Earth is not a living organism, but as a productive program of environmental research and a spur to political and moral action, Gaia lives.

10

From Egg to Organism

Except for the workings of the brain, no other phenomenon in the living world is as miraculous and awe-inspiring as the development of a new adult from a fertilized egg.

—Ernst Mayr, *This Is Biology*[1]

How an egg turns into an animal has challenged the imagination of naturalists since antiquity; Aristotle was probably the first to write a treatise on the subject. Two millennia later we continue to marvel at the purposeful manner in which a featureless egg produces a fruit fly, a sea urchin, or a baby, each according to its own kind; and each made up of millions, even trillions, of precisely specified cells arranged into a coherent body that persists, maintains itself, and starts another round of the same cycle. The performance raises a host of questions about how it all works and how it came to be that way, most of which have no more than partial answers. All the same, in broad outline, development no longer seems mysterious or miraculous so much as marvelous. Some complain that science has taken the enchantment out of watching a caterpillar turn into a butterfly, that delicious sense of mystery that pervaded the natural world when we were young. Scientists, of course, do not see it so. We are proud of our effort to make rational sense of nature, and are keenly aware that we have only substituted one wonder for another: How can an ensemble of lifeless chemicals build a butterfly?

A Long Road to Discovery

The contemporary understanding of development (and of what that word means) grew slowly over centuries. An early landmark, in the 17th century, was the work of William Harvey, who cracked open chicken eggs on consecutive days of the three-week incubation period, and described what he could make out with a hand-lens. Later milestones include cells, nuclei, the dance of chromosomes during cell division, and the vehicles of reproduction, or gametes—the egg and the sperm. Key questions in the early years were the nature of fertilization and the extent to which male and female contribute to the emerging embryo. Not until well into the 19th century was it generally recognized that both parents contribute equally, that the essence of fertilization is the union of a sperm with an egg to produce a zygote, and that at bottom sperm, egg, and zygote each consist of but a single cell.

Conceptual debates centered on how the formless stuff of a fertilized egg could bring forth a frog or a fish, with proponents of two conflicting visions vying for authority. One school, the "preformationists," contended that miniature versions of the future adult were already present in the gametes and needed only to be unfolded. This implausible notion is epitomized by the image of a tiny homunculus scrunched up in the head of a sperm. The opposing school, the "epigeneticists," argued instead that embryonic forms are produced in the course of development, under the guidance of an intrinsic but wholly mysterious directing force. This dichotomy pervaded the study of embryology well into the 19th century, not to be fully resolved until the rise of genetics transcended both positions. It is clear now that novel structures do arise in the course of development, and no homunculus lurks in the sperm. The force that directs development is built into the gametes in the form of their complement of genes, some of which are dedicated to development. Genes are what ensures that dogs produce puppies and never kittens, while the converse is true of cats. The set of developmental genes is commonly said to hold a "genetic program" of development; misleading though this expression often is, it is almost unavoidable.

The heyday of embryological research came in the latter 19th and early 20th centuries, supported by ever more powerful microscopes and the use of surgical procedures to query the causes of development. Among its achievements was the demonstration that the early stages of animal development entail extensive cell migration. These highly orchestrated movements generate the three characteristic "germ layers," which eventually give rise to all the tissues and organs. The fertilized egg holds the potential to form any and all the cell types of the adult, but as development proceeds the cells produced become increasingly committed: some will mature into liver cells, others into brain or muscle, and in the normal course of events that decision is irrevocable. The nature of the various cells is defined in the first instance by the proteins they contain, products of the differential expression of their genes. That, in turn, is dictated partly by their own nature and partly by signals from neighboring cells in the presumptive tissue. Just how cell differentiation comes about is one of the central issues in contemporary research.

Unlike the basic operations of cell biology, development is a multifarious phenomenon. We saw in Chapters 2 to 5 that there is in principle only a single way to translate genetic instructions into proteins, or to generate useful energy (with intriguing but modest variations). By contrast, each embryo develops in its own fashion and the underlying unities are hard to discern. Both animals and plants operate by way of embryos, but shape them in entirely different ways. Rather than describe particular pathways, let me in what follows focus on three of the general problems that all developmental trajectories must solve: differentiation, pattern formation, and morphogenesis.

Going with the Program

Like other potent metaphors, the "genetic program" illuminates reality, but also distorts it. Suppose it were possible to carry out the Jurassic Park experiment: insert a dinosaur nucleus into an enucleated crocodile egg and have it hatch. It's a safe bet that the hatchling would be a dinosaur of sorts, not a crocodile, reinforcing the general understanding that the specific information required to make an organism is contained in the nucleus, not the cytoplasm. Moreover, the pertinent genes make up a subset of the genome: some 1,500 genes out of a total of about 15,000 in the fruit fly *Drosophila*, perhaps more in mammals, are chiefly or wholly dedicated to development and could be described as encoding a program. But matters are more complicated than that. We saw in Chapter 4 that both genes and cellular organization are required to make a cell, and this is doubly true of the development of an embryo. The choice of which genes to activate, and also where and when, depends on the state of the developing embryo as well as signals from neighboring cells. Unlike the program of a theatrical performance, the program of development cannot be separated from its execution; genome and cell structure are partners, albeit somewhat unequal partners. If you find the metaphor of a program informative feel free to cling to it, but do not imagine the genome as a blueprint. Better think of it as a recipe, as in baking a cake (Richard Dawkins), or as the instructions for folding a sheet of paper into origami (Denis Noble).

In former days it was widely believed that, as the zygote undergoes differentiation into liver, muscle, or brain, it jettisons genes that are no longer required. That turned out not to be the case: tissue cells retain the zygote's full gene complement, and can in principle (sometimes even in practice) revert to their undifferentiated state. Instead, cell differentiation relies on the selective control of gene expression. Some genes are permanently silenced, commonly by chemical modifications that do not involve the nucleotide sequence ("epigenetic regulation"). Others are induced, repressed, or modulated in a sophisticated pattern that begins to draft the organization of the embryo. These genes typically encode "transcription factors," proteins that regulate the expression of other genes.

We have known for half a century that one level of control in cell metabolism relies on proteins, specified by regulatory genes, that bind to particular sites on DNA and promote or repress the expression of the genes that encode metabolic enzymes. Developing embryos carry this principle to a higher level: the expression of genes that specify the working proteins (cytoskeleton, say, or signals) is controlled by a cascade of regulatory genes arrayed in hierarchical networks that look not unlike an old-fashioned telephone exchange[2] (Figure 10.1). Transcription of some gene high up in the hierarchy produces

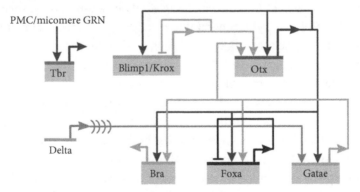

Figure 10.1 Part of the gene-regulatory network involved in the specification of sea urchin mesoderm. From Davidson and Erwin 2001, with permission of D. Erwin and the American Association for the Advancement of Science.

a transcription factor that alters the output of another gene lower down on the pecking order. That second-tier gene will likely receive input (in the form of additional transcription factors) from several other sources as well, so that its output represents something like a consensus. Genes of the second tier communicate with a third, still lower down, which may control the working gene. The operation of such gene regulatory networks, several of which were mapped in impressive detail by the late Eric Davidson and his colleagues, cannot be apprehended by the unaided intuition but requires computational analysis. The upshot is that hierarchical networks of control explain many of the features of development that seem so magical: the unambiguous speci-fication of cell fate (it's either liver or brain, never anything in between); the inexorable progression of events from egg to embryo to adult; the precise definition of regional boundaries; and robustness in the face of perturbation or error.

Pattern Formation

How can an undifferentiated mass of cells know which ones must activate the proper gene cascade, so as to produce an embryo with all its parts in their proper place? That is the problem of pattern formation, and it takes us one level beyond the molecular. The fruit fly *Drosophila* has proven to be singu-larly convenient, both because it is genetically tractable and because spec-ification of cell fate takes place even before the formation of discrete cells. Numerous mutant lines have been created, whose progeny display abnor-malities of form and structure. The genes involved in shaping the fly are commonly designated by whimsical names derived from the mutant pheno-type: *hunchback*, *sevenless* and *bride of sevenless*, *hedgehog*, and *gurken*, and their abnormalities proved immensely instructive.

Very briefly, the early *Drosophila* embryo develops along two spatial axes: one anteroposterior (front to rear), the other dorsoventral (back to belly). These axes are initially specified in the follicle that houses the fertil-ized egg, by messenger RNA produced by the mother fly. Translation of that RNA yields molecules of a protein that diffuse away from their site of produc-tion, forming a gradient with a range of several cell diameters that supplies positional information to cells under its sway (Figure 10.2). Such molecules (often but not always proteins) are designated "morphogens": the concentra-tion of morphogens at any point in the gradient instructs the recipient cell

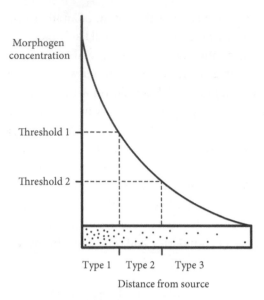

Figure 10.2 A morphogen gradient conveys positional information. Cells exposed to a morphogen concentration above threshold 1 follow one course, those exposed to concentrations between 1 and 2 another, those exposed to the lowest level (or to none at all) follow a third.

as to the course it should follow. For example, in the case of the fruit fly embryo the gradient of the maternal protein Bicoid (maximal in the future head region) maps out the anteroposterior axis. Wherever the concentration of Bicoid exceeds a threshold level it turns on a gene of the embryo, *hunchback*, whose product is another transcription factor. This, in turn, governs genes further downstream that specify the identity of embryonic segments.[3]

When the regular interplay of genes, cells, and signals is disrupted by mutation, the outcome can be startling. A case in point is the gene *pax6*, whose expression is pivotal to the formation of *Drosophila*'s eyes. Normally, of course, eyes form in the head. But it is experimentally possible to activate *pax6* in other tissues, and induce them to bring forth eyes—on the legs, under the wings, and so on. Not only will the *pax6* protein of *Drosophila* do that but so can its mammalian homolog, demonstrating the degree of molecular conservation over the 600 million years since insects and mammals diverged from a common ancestor. It seems that the product of *pax6* is a master transcriptional regulator, a general switch not unlike the hallway switch in your home which, when flipped, turns on the lights and the heater also.

Another bizarre mutation affects the location of organs in the segments of the fly. In the normal course of events, the head is made up of the first three segments, one of which bears antennas. Mutants defective in the gene *antennapedia* develop quite normally, except that instead of antennas the head sprouts a pair of legs! It seems that the segment in question has switched identity, from head to thorax. This grotesque transformation was the clue that led to the discovery of the *hox* genes, a set of genes instrumental in specifying the identity of the fly's body segments. The genes are arrayed in linear order which corresponds to that of the segments. Homologous genes occur in other animals, including vertebrates, always in the same order as the parts of the body. Just what this correspondence means is uncertain, but it again illustrates the deep and highly conserved commonalities that underlie the conspicuous diversity of living things.

To Shape an Embryo

Genes are linear, one-dimensional sequences of nucleotides. Even a morphogen gradient has but a single vector. By contrast, an embryo is intricately formed in three dimensions, the adult even more so. How bodies are generated is the subject called morphogenesis, which takes us a very long way beyond the molecular level. Genes do not construct bodies, only cells do that. Genes inhabit the nanometer scale of sizes; cells, 10 or more micrometers in length, are a billion times larger by volume. Multicellular organisms, made up of myriads of cells connected into tissues, carry us into the range of millimeters to meters, larger still by many orders of magnitude. It follows that what organisms do is separated from the genes that specify their molecular parts by multiple layers of complexity, physiology, and diversity.

Animals, plants, and fungi must all meet the challenge of constructing large integrated bodies from cell-sized building blocks, and therefore the most basic aspects of morphogenesis are common to them all. One such feature is that living forms are products of growth: the multiplication of cells, often also their enlargement. Another is that eukaryotic cells are polarized: one end differs from the other, both in structure and in functions. Morphogenesis is conspicuously dependent on the physical forces that shape and reshape cells, tissues, and entire organisms. These operate in conjunction with chemical affinities and reactions at cell surfaces. Regulatory gene networks and chemical messengers turn up everywhere. Nonetheless, it has proven very difficult to discern laws that govern what shapes are permissible to living things, and there are profound differences in the way organisms shape themselves.

The Animal Way

Embryogenesis in animals turns on three characteristic processes: cell adhesion, cell and tissue migration, and the deployment of mechanical forces. Adhesion between cells of the same kind, mediated by a class of surface proteins called cadherins, is what holds tissues together. That became clear from experiments first conducted half a century ago, in which embryonic tissues were gently disaggregated into their constituent cells, which were then mixed and allowed to sort themselves out. As the cells reaggregated, like bonded with like: presumptive epidermal cells joined into one "tissue," neural precursors into another, with the epidermal cells on the outside and the neural cells interior. Experiments of this kind supplied the basis for the isolation of the proteins whose mutual affinity underlies the selective aggregation of cells into tissues.

As noted earlier, animal cells are motile and their shapes are flexible. Division of the zygote usually produces a hollow ball of cells called a "blastula." This then embarks on an orchestrated set of movements, in which a particular subset of cells loosen the bonds that tie them to their neighbors and migrate into the interior of the blastula (Figure 10.3). "Gastrulation," as this reorganization is called, gives rise to the three concentric germ layers that generate all future tissues: "ectoderm" (skin, also the neural system); "endoderm" (most of the viscera); and "mesoderm" (musculature, gonads, blood). As Lewis Wolpert once observed, the most important event in your life is not birth, death, or marriage—it's gastrulation.

Shape changes, together with cell migration, are the chief drivers of animal morphogenesis. A well-studied example is the manner in which sheets of cells fold: the key is that cells within the presumptive fold undergo contraction at the apex. This contraction forces the cells into the shape of a wedge, and since all the cells are bonded together it makes the sheet bend. We have here a fine example of the deployment of mechanical force, generated at the cellular level, to do work at the level of the organism. Force is generated by the cytoskeleton, the requisite energy comes from the expenditure of the chemical energy currency ATP (Chapter 3), and localization is effected by

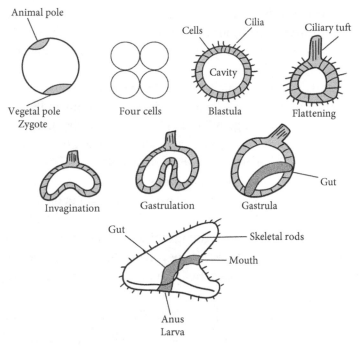

Figure 10.3 Early stages in the development of the sea urchin. Inspired by Gilbert 1991.

chemical signals. Much the same holds for cell movements, which again rely on contractile forces generated by the cytoskeleton, guided by chemical or electrical signals.

Embryogenesis is an intensely active area of research, with an enormous specialized literature. It holds detailed descriptions of the course of events in numerous organisms, and the underlying mechanisms are increasingly well known. But it is not obvious what general principles explain how the shapes of fly, frog, or sea urchin emerge from these particulars. Are there none, or does something fundamental continue to elude us?

Plants Do It Differently

Plant cells are enclosed in stiff walls, they are not motile, nor are they easily deformed. Development of a plant zygote, like that of animals, revolves around differential gene expression, chemical communication between cells and positional information, but cell migration is not involved. Instead, the organism's form emerges by localized growth that calls for the multiplication of cells and their enlargement, with attendant changes in ambient mechanical forces.[4]

An animal at birth is essentially a miniature adult. Limbs and organs are in place, and subsequent development affects proportions and functionality. By contrast, in plants the germling displays but a sketchy outline of the future adult, with all the organs generated after the plant has begun to grow. This is illustrated in Figure 10.4, which depicts stages in the outgrowth of the weed *Arabidopsis thaliana*, today the favorite research organism for plant biologists. The zygote begins to take shape in the ovule, but then remains dormant until the seed germinates. That event initiates an orchestrated sequence of divisions that generate the seedling with its rudiments of the adult plant-to-be: cotyledons (future leaves), hypocotyl (future shoot), and a root. All these organs derive from two permanent growth centers, called meristems,

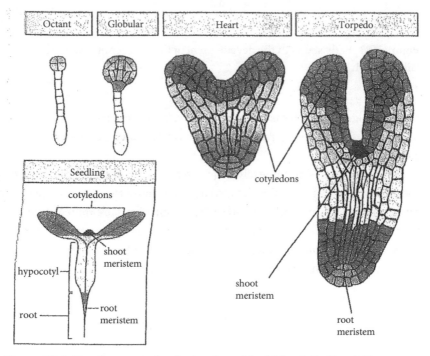

Figure 10.4 Development of a plant embryo (*Arabidopsis thaliana*). From Wolpert 2011, with permission of Oxford University Press.

where new cells arise by continuous division. One daughter undergoes differentiation, the other remains in the meristem and goes on dividing.

How do plants determine which particular cells are destined to differentiate into leaves, roots, and eventually flowers? In animals, patterning of the embryo is effected chiefly by diffusible morphogens that instruct cells as to their position on the fate map. Chemical messengers, plant hormones, are also part of the developmental strategy of plants, most notably the ubiquitous auxin and cytokinin. However, as a rule these do not function as diffusible morphogens. Auxin, for example, is actively transported from one cell to another. Moreover, in plant tissues adjacent cells are commonly linked by structural channels that allow molecules (including proteins) to bypass the extracellular space. Mechanical forces demonstrably play a major role, but just how regional patterns arise in plants is not clearly understood.

We are rather better informed concerning the expansion of plant cell walls and its localization. The heart of the matter can be summed up in the phrase, "global force and localized compliance."[5] What drives expansion is hydrostatic pressure, commonly called turgor, which results from the tendency of water to flow passively into plant cells in response to their accumulation of metabolites and ions. The rigid cell wall resists the influx, generating the pressure that keeps plants upright (that's why your houseplants wilt if you forget to water them). Hydrostatic pressure acts uniformly at all points; were the wall to yield equally everywhere, the result would be spherical expansion (Figure 10.5). In fact, mechanisms are in place to localize expansion, thus producing more asymmetric forms. Extensive research over decades has highlighted the roles of localized acidification of the wall (effected by the cytoskeleton), and a special class of enzymes called "expansins" that make cellulose fibrils in the acidified region come apart, allowing the wall to yield locally to the global force of turgor.

It would be satisfying to conclude this section with a concise and coherent account of how plants shape, say, a leaf; but we are not there yet. We can watch the unfolding of the zygote into a seedling, and eventually into an adult; and one by one the players are coming into view. But it is still not possible to make out how interactions among the molecular players make the leaf.

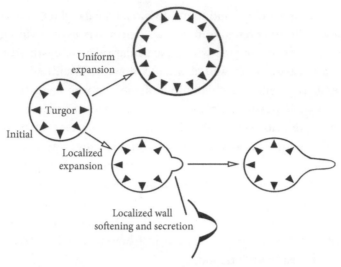

Figure 10.5 Global force and local compliance. Hydrostatic pressure (turgor) acts uniformly in all directions, and drives spherical expansion. Nonuniform enlargement, such as the emergence of a root-hair, requires localized compliance with the global force, mediated by the cytoskeleton. Inspired by Mathur 2006.

How Hyphae Grow

Fungi are not prominent in the literature of morphogenesis, and indeed the manner in which large and complex mushrooms come into existence is understood only in general terms. But at a lower level of size, how fungal hyphae grow at the tip is increasingly comprehensible and can even be described mathematically. If ever it becomes possible to "compute an organism," hyphae may be candidates of choice (Figure 10.6).

The hypha is the basic unit of fungal growth and form, a filament of interconnected cells large enough to see with the naked eye. The white fungal matter that your trowel turns up in garden soil is a mass of hyphae. Their distinguishing feature is "apical growth": Extension of a hypha takes place exclusively at its tip, or apex. To accomplish this, hyphae mobilize resources (building blocks, enzymes, energy, membranes) from all along the trunk, package them into vesicles and transport them to the apex. There they are secreted, the vesicle membranes contributing new apical membrane while the enzymes build fresh wall. The driving force for extension is (usually)

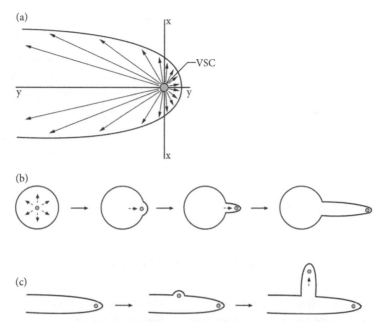

Figure 10.6 How fungal hyphae grow. (A) The vesicle supply center. The "hyphoid" is an idealized curve that corresponds very well to the shape of living hyphae, calculated on the premise that secretory vesicles emanate from a forward-moving source. (B) Displacement of the vesicle supply center from a central position generates a germ-tube. (C) Formation of a lateral branch by the initiation of a new vesicle supply center. From Riquelme et al. 2018, with permission of Dr. Riquelme and the American Society for Microbiology.

turgor pressure. Well-fed hyphae put forth branches from the trunk, each one led by a new growing tip.

The locus of hyphal morphogenesis is the very apex, where spatial organization is continuously created anew. The key to how that comes about is the "Spitzenkoerper" (German for "apical body") an ephemeral structure visible only as long as hyphae are extending. After considerable controversy, its role now seems solidly established: a dynamic, self-organized structure produced by the flow of vesicles, which coordinates the final stage of their journey to the tip. From this simple premise, Salomon Bartnicki-Garcia and his colleagues were able to derive a mathematical formula that correctly predicts the characteristic form of a hyphal tip, called a "hyphoid"; an achievement doubly notable for its rarity in the annals of morphogenesis.[6]

Beyond this point matters become complicated, as the molecular mechanisms are progressively being clarified. Fungal cell walls, like those of plants, are rigid; what makes them plastic at the tip and then allows them to harden? How many kinds of vesicles are carried forward, and what do they contain? How does the cytoskeleton support vesicle transport and shape the tip? There are many questions and quite a few partial answers; these topics fall outside the scope of this book. Also, and to my regret, I can mention only in passing the penetrating studies of Peter Hepler and colleagues on apical growth in pollen tubes.[7] But the sense of progress is palpable: one can begin to see how systems physiology mediates between the molecular players and the morphogenetic plot.

The Evolution of Forms

Multicellular organisms normally begin as single cells, and can only get to the adult stage by way of development. The diversity and progression of living forms must therefore be due, at least in large part, to variation and selection acting at the level of development. How did that come about? The key discovery is that genes exercise their developmental role as members of a system, a regulatory network embedded within the larger system of cell physiology. This viewpoint allows us to transcend the stale argument over whether genes make up a "program" of development, to focus on what they actually do and how these informational networks evolved.

Let's begin with a long-standing puzzle. All the body-plans of animals (the phyla) made their debut in the Cambrian era or shortly before. They have endured for some 500 million years, but in all that time no new body-plans have arisen; why would that be so? It's not that evolution ran out of steam: innumerable innovations appeared within the various phyla, including four legs to walk on; the wings of insects, birds, and bats; immune systems; digestive enzymes and structures; and so on. Are the known body plans the only possible ones? Are they products of numerous small changes that eventually produced major adaptive transformations? Eric Davidson, whom we met in a previous section, offers a more plausible explanation based on the principle that development is controlled by gene networks, arranged hierarchically so that genes higher up modulate the expression of those lower down.[8]

At the top of the pecking order we find "kernels," regulatory circuits such as the one diagrammed in Figure 10.1, that were established very early in animal evolution. The genes that make up those modules mostly encode transcription factors that map out the body's regions, and each impinges on many genes below. Assembly of the kernels underlies emergence of the largest groupings; for example, the Bilateria, all animals from flies to earthworms and humans, that share bilateral symmetry. Because of their critical functions, in those genes mutations cause massive disruption of development, likely to end in their bearers' early death. Some, indeed, appear to have been scarcely altered since the Cambrian. The tier below consists of "plug-ins," circuits used repeatedly to control what is produced by any particular region. Many morphogens fall into this class. Plug-ins receive input from the kernels, and in turn regulate the genes that execute the program. On the evolutionary timescale, genes of this class are more labile than those of the kernels, and they account for the bulk of morphological evolution. The lowest tier consists of gene batteries that produce actual structures—muscles, bones, and nerves. These make up the network's periphery: they are regulated by genes higher up, but do not regulate others. This is the species level of evolution, the one least constrained because the consequences of change are limited in scope.

From this perspective, genetic changes are not all alike. Change at the kernel level is likely to be catastrophic, and seldom happens. Once assembled, kernels are not lightly disassembled, they can only be built upon. Moreover, once the existing patterns were established they altered the environment in ways conducive to their own persistence. This is why there have been no new phyla in half a billion years. The bulk of evolution takes place in the lower tiers, creating the stunning diversity of living forms built on a narrow palette of body plans.

Sean Carroll[9] has done the field great service by formulating the outlines of a general theory. In the classical view, the emphasis is on genes that encode working proteins, evolving by small steps guided by selection for improved function. It was therefore expected that organisms which differ in form would also differ in their functional proteins. Surprisingly, that proved not to be the case. Proteins from humans and chimpanzees, for instance, are mostly very similar; and what variation exists appears to be due to genetic drift rather than adaptation. The expectation that gene duplication would be a major mechanism of protein evolution has also not been confirmed. Instead, most of morphological evolution is due to changes in regulation, particularly within the networks that control the expression of the genes that encode working functions. As a result, major changes can take place in one or a few steps.

It is network architecture that explains why the evolution of form is so highly constrained that ancient kinships remain visible even after hundreds of millions of years. The basic toolkits are conserved because network architecture is interactive rather than modular; animals that diverged long ago still rely on their original tracks and switches. Thus an animal *pax6* gene can take the place of the fruit fly gene, even though the eyes of insects and animals could hardly be more different and originated independently. The proteins that regulate development participate in multiple distinct processes, and show up repeatedly in the patterning of body plans. Network topology tends to be preserved. This, I suppose, is also why the development of embryos sometimes recapitulates features of earlier stages, such as the transient appearance of gill-slits in human embryos.

In writing of these matters I have found it impossible to avoid the language of a gene-centered view which implies that genes specify, regulate, and control bodies. This is presently the common understanding, and it dominates the public conversation of biology. It is high time to reexamine that basic premise in the context of development, and clarify how genes do fit into the baffling scheme by which an egg turns into an organism.

Is Anyone in Charge Here?

To my mind, what is most baffling about development is its purposeful nature, the single-minded discipline that rules the proceedings and ensures that they lead almost infallibly from egg to fruit fly, sea urchin, or baby. We are, of course, generally familiar with the production of complicated objects like airplanes, and we understand how that is done. The key is a plan, a blueprint, that gives direction to the whole enterprise and marshals innumerable individual steps to the final objective. Is there anything equivalent to guide development? Molecular scientists have traditionally and loudly said Yes, pointing to the program encoded in the genome. In its contemporary form this belief goes back to the heyday of molecular biology sixty years ago, and its iconic formulation was given by the late Francois Jacob: "The whole plan of growth, the whole series of operations to be carried out, the order and the site of synthesis and their coordination are all written down in the genetic message."[10] I hasten to add that Jacob's views soon became more nuanced but the preceding proclamation is the one that stuck, and it represents an attitude still widely held by scientists and the general public alike.

It is now clear that the relationship between genes and development is much less straightforward, and Jacob would surely not stand by his sweeping claim today. In a nutshell, we understand that genes specify molecular chemistry: directly so for the sequences of RNA and proteins, indirectly and more loosely for other biomolecules. Genetic loci are also heavily involved in controlling the expression of coding genes, reaching an apogee of sophistication in gene-regulatory networks. But genes, either singly or in networks, do not specify higher levels of biological organization: the location and spatial direction of events, or the forms of cells, organs, and organisms. To be sure, these also are ultimately grounded in molecular constructs, and so genes can never be left out of account. But there seems to be no way to encode three-dimensional order in linear sequences; you need cells for that. We saw this in Chapter 4 in the context of cell growth and morphogenesis, and it is doubly true of multicellular organisms because the difference in scale between genes and organisms is that much greater. There are not nearly enough genes in a genome to specify every aspect of the organism. Moreover there is abundant evidence that the higher levels of organization routinely inform, constrain, and guide the expression of the genetic database (albeit not usually its content). For example, mechanical forces have been shown to elicit the activation of previously quiescent genes, thus completing the circle from genes to organisms and back again. Genes do not hold a dominant blueprint, they operate in the context of cells, tissues, and organisms. Parallels can be drawn to recipes, musical scores, origami, or even painting, but in the end all of them fall short. Living systems are in a class of their own, and there seems to be no satisfactory analog for a system that makes itself from parts specified by the system itself.

We arrive, then, at the cloud of unknowing that hangs over all of developmental biology: How can organized systems, on the scale of millimeters to meters, govern their own formation without any discernible plan? In all my reading over the years I have never come across a fully satisfactory answer. The general principle, it appears, is the somewhat nebulous phenomenon of self-organization; it is unquestionably real, but hard to grasp. Self-organization is the process by which, in complex systems subject to the flow of energy, large-scale order emerges on its own from the interaction of elementary parts that obey only local rules. Examples range all across the spectrum from physics to sociology. They include the pattern of hexagonal convection cells that form in a pan of liquid heated from below; the regular arrays of rocks sometimes encountered by travelers on arctic and alpine tundra; the spontaneous appearance of dynamic structures when microtubules are incubated with motor proteins and an energy source; the coordinated movements of flocks of birds and schools of fish; even the gyrations of the stock market.

The progressive transformation of an egg into an organism is best envisaged as a grand instance of self-organization. There seem to be no general maps to guide the cell migrations that produce the gastrula and early embryo in Figure 10.3, or the pattern of oriented and localized cell divisions that shape the leaf in Figure 10.4. Each molecule, every cell, responds only to local circumstances; they do not "know" that they are actors in a greater play, yet order on a far larger scale emerges spontaneously. No one is in charge beyond the dynamics of the system itself, a web of activities that takes the same course time and again.

Some strands of that web can be made out. The most comprehensible is the rough grid laid out by morphogens, which sketches a broad spatial map. More subtle, and much less studied, is the net of stresses, strains, and mechanical forces that links a host of individual cells into an integrated whole. Cells adhere to one another at specialized foci, where forces generated by the cytoskeleton are transmitted to neighboring cells. At a higher level, connections knit together bones and muscles, such that a minor change in posture may relieve a pain in the neck. Principles of tensegrity, the architectural basis of Buckminster Fuller's famous domes, supply a model. Whether those linkages are sufficient, and what might supplement them, remains to be worked out. By doing that we can hope to gain better insight into how system dynamics produce the integrity, the wholeness, of a living organism. When that puzzle yields, answers will likely take the form of models that integrate inputs from many sources; they will be intelligible to a computer, but perhaps not to the unaided intuition.

In the meantime, anyone seeking a broad-gauge understanding of development must be content with metaphors.[11] I know of none more profound than the "epigenetic landscape" evoked by Conrad Waddington sixty years ago (Figure 10.7). Waddington likened development to the path of a ball rolling downhill through a tilted landscape of valleys, to a predictable

(a)

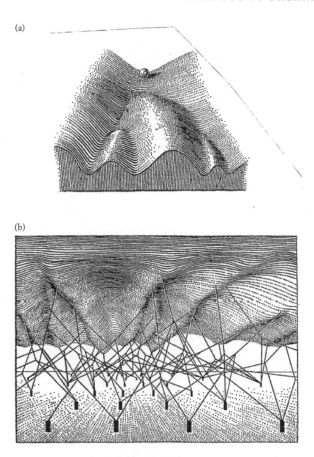

(b)

Figure 10.7 The epigenetic landscape. (A) The course of morphogenesis is represented by the path of a ball as it rolls downhill toward the observer. Normally the ball will keep to its left, but it may be diverted into an alternative channel by genetic or environmental change. (B) System of interactions underlying the epigenetic landscape. The pegs in the ground represent genes, and the strings leading from them correspond to the activities of their products. Reprinted from Waddington 1957, with permission of Taylor and Francis Informa UK Ltd.

endpoint at the bottom. In the normal course of events the ball will keep to the (reader's) left at the fork, but it would not take much to push it over the low divide into the adjacent valley, and an abnormal endpoint. So, how is that landscape formed? Look at it from beneath, and note that it is a canvas whose hills and dales are shaped by ropes attached to pegs in the ground. The pegs symbolize the genes, the developmental program. They do not specify particular features but sculpt the landscape in which development unfolds. And genes are not the only factor: cellular and environmental factors influence the tension on the ropes, so that both genetic and physiological inputs collaborate to make the landscape.

11

The Outer Banks of Order

There once was a man named Descartes
Who thought he was terribly smart
Saying matter and mind are of opposite kind
Thus keeping them ever apart.

—Keith Chandler, *The Mind Paradigm*[1]

Let us now step back and contemplate the history of life whole, all 4 billion years of it, and ask once more whether it is just one damn thing after another or holds a discernible storyline. The answer seems obvious and has been for over a century: despite all the complications, contingencies, diversions, and extinctions, there clearly is a trend toward mounting complexity, organization, and sophistication. The tale begins, or so we believe, with chemical systems about which we know next to nothing, and by around 4 billion years ago features the first rudimentary cells. Both the fossil record and molecular phylogeny indicate (at least, as most of us presently read them) that, for the first 2 billion years of life's history, all the life that lived was of prokaryotic grade and that Bacteria and Archaea were both present. Eukaryotic cells make their first appearance much later, 1.6 to 1.8 billion years ago, and ever since they have made most of the evolutionary running. Prokaryotes were, and are, biochemically versatile but have remained small, structurally simple, and fundamentally unchanged. The diversity of microbial eukaryotes exploded some 1.2 billion years ago, and multicellular creatures make their debut some

600 million years ago. They grew in size and powers all through the Paleozoic era; by the Triassic, 200 million years ago, forests covered the land, dinosaurs roamed the swamps, and mammals were on the horizon.

What next for the gyre of complexity? Ever since remotest antiquity it has been taken to be self-evident that we humans are special thanks to our superior minds. This premise, shorn of its traditional religious connotations, still seems to me perfectly sound but has gone out of fashion. Scientists like to minimize the exceptional status of humans on the grounds that human physiology puts us squarely among the animals, and also that many animals possess mental and emotional faculties. That is all quite true, and so is the mounting consensus that mind is a product of evolution by heredity, variation, and natural selection. My point is rather that the advent of mentation, early in the rise of animals and possibly long before, seeded the emergence of higher levels of rationality and will that are now transforming the entire globe—for good or for ill. Geologists and biologists speak of our time as the Anthropocene, an era that began with the advent of farming some 10,000 years ago, and can be recognized by signs of the environment being significantly impacted by human activities.

So the subject of this chapter is mind, its nature and origin, with an unavoidable emphasis on human minds and what they have wrought. Whether we should celebrate the attainments and promise of our singular species, or dread the consequences of our relentless quest for power, still remains to be seen. At present the omens are grim, but on the long scale of evolution even 10,000 years are but a moment; and what the end will be no one can tell for certain.

Mind from Matter

The enduring conundrum of mind versus matter was first posed by the French philosopher René Descartes (1596–1650), though that may not have been his intention.[2] Writing at a time when the Catholic church still exerted much baleful power over philosophical inquiries, Descartes sought to open space for the rational exploration of nature while minimizing conflict with the authorities. He was well aware of what had happened to Galileo; Descartes himself eventually left France for Holland, and later settled at the court of Queen Christina in Stockholm for the remainder of his life. In his scheme the proper study of science was the material world, while mind and soul stood altogether apart; and such has been Descartes's influence that the bright line between mind and matter endures to this day.

The dichotomy is certainly not without substance, and remains unbridged. "Mind," which Merriam Webster's Dictionary defines as "The element or complex of elements that feels, perceives, thinks, wills and especially reasons," is obviously not wholly a material phenomenon. There is nothing in the vocabulary of physiology or biochemistry, let alone physics or chemistry, that connects with mental activities such as memory, affection, judgment, or morality. But neither is mind altogether divorced from matter, as when an immaterial "I" decides arbitrarily to lift a material finger, and its physical locus is intimately linked to nerves and brains. Moreover mind, like all other features of biology, is a product of evolution and that is a good place to begin to reflect on its place in the natural world.

The proposition that mind evolved over time, in tandem with the material brain, has been a sticking point for many people and featured in the 19th-century debates over evolution. Humans have minds, no one disputes that. But do dogs, earthworms, or plants? As we cannot ask them about their interior lives, we must judge by their behavior; and those who live and work with animals do not doubt that their companions have attributes that we credit to mind, including intelligence, anxiety, affection, desire, and memory. Darwin himself saw clearly that the minds of humans and animals differ, but the difference is one of degree rather than of kind.

Matters become more ambiguous with "lower" creatures. Do snails have minds? The marine snail *Aplysia* learns to associate certain stimuli with a subsequent electric shock, and to take evasive action. It displays a degree of memory, and has become an important subject for research in neuro-physiology. Cephalopods, including octopi and cuttlefish, have an extensive behavioral repertoire. They are often very clever and seem to be aware of themselves, to the point of appearing conscious. Recent studies suggest that bees are far from being mental zombies acting out instinctive routines in a mechanical fashion. They can learn, and may be moody. Protozoa often actively pursue their prey and exhibit some kind of memory; even bacteria have a (very short) attention span. I see no basis for doubting that minds, including our own, have evolved over time, presumably by the usual interplay of heredity, variation and natural selection. It is a phenomenon whose roots go clear back to unicellular forms,[3] and is grounded in the material laws of chemistry and physics. The sophisticated minds we reflect with today are confined to multicellular organisms, and are indissolubly linked to neural processes; generally speaking, mind increases in parallel with the size of the brain.

We saw in previous chapters that evolution is the result of natural selection acting on changes at the level of the genes, and also at the epigenetic or physiological ones. So, are complex patterns of behavior somehow encoded in the genome? The physical infrastructure certainly is, for it is built of gene-specified elements: all those cells, neurons, synapses, and neurotransmitters. In some cases particular mental operations have a genetic basis. A case in point is the gene *fruitless*, which is required for courtship behavior in *Drosophila*: the expression of a functional *fru* genes is both necessary and sufficient for males to strut their stuff.[4] In humans the capacity to speak and learn language is inherited, and a few genes have been identified that are required for the language instinct to be activated, but whether you learn to speak Swahili or Slovenian has nothing to do with your genome. These particulars grow out of a much higher level of biological organization, at which cells (neurons) rather than genes form elaborate communicating networks.

An elementary network of neurons underpins the behavior of the marine snail *Aplysia* mentioned earlier, the proud possessor of some 20,000 neurons (compared to billions in higher organisms). For purposes of research a special virtue of *Aplysia* is that its neurons are exceptionally large, making it easy to record electrical impulses with intracellular microelectrodes. The snails learn to associate a gentle touch with a later electric shock, and withdraw their mantle in anticipation. The memory persists for hours, in some situations longer. What is the physical basis of that memory? Though much remains to be discovered, there is substantial agreement that memory resides in patterns of particular neurons that become active together when stimulated, and retain their linkage through the dendrites that connect one neuron to another. As they say in the trade, neurons that fire together, wire together. Neuronal networks in more advanced organisms are correspondingly more elaborate than in snails, but the basic principle of an "engram," a cluster of neurons physically coupled via their dendrites, appears to hold.[5]

If mind, over all its levels from simple behavioral reflexes to our own sentience, is a product of Darwinian evolution, what was it selected *for*? There is no mystery about patterns of behavior such as fighting, fleeing, or mating: these are immediately useful to their possessor and are likely to have been preserved and honed over time. As Daniel Dennett says, they are good moves in design space. Many higher levels of perception, cognition, memory, and learning are likewise of obvious benefit. To quote Dennett once more, "Thinking ahead is the main task of nervous systems: preparing the organism to make timely responses to its multifarious needs and opportunities."[6] But what of consciousness, or self-awareness (Henry Gee)—the quality of being aware of one's own internal and external state? Would abstract abilities such as musical or mathematical talent be favored by natural selection? Perhaps, or perhaps not; such skills may also be gratuitous expressions of the intricate networks that underlie all mental operations, not directly selected in their own right but doubly precious for all that. For here we glimpse the foundations of civilized societies, and touch qualities that have given humans dominion over all the earth.

So far so good, but the argument that mind is grounded in matter and a product of evolution sidesteps the hard problem: What *is* mind, what is it that makes a passing scent call up memories going back decades, and underlies my absolute certainty that I am a real person who inhabits (even controls) a body? Who or what is it that commands *me* to lift *my* finger? Granted that these everyday marvels are expressions of neuronal networks of surpassing intricacy (the human brain, a three-pound lump of matter with the consistency of tofu, is said to hold some 100 billion neurons), how do sensations, memories, self-awareness, and thought emerge from all those myriads of firings and couplings?

We really have no answer that satisfies the ancient Greek injunction, Know Thyself.

Biologists are apt to rank sentience as one of the two deep mysteries that bookend our science: the origin of life at one end, consciousness at the other. But there does seem to be some convergence on the point of view vigorously articulated by Daniel Dennett, that consciousness is not a special faculty that some brains have mastered. Rather, it simply *is* the workings of that mind-bogglingly complex network of neurons, and will be found (at least in rudimentary form) wherever neurons converse.[7] From Dennett's standpoint it is meaningless to ask how consciousness evolved or what function it serves, for consciousness as such does not exist. It is not a specific faculty separable from the mundane operations of the brain. There is no mental theater in which I sit, viewing the action and manipulating events. Instead, perceptions and experiences are continuously received, massively processed, mulled, and massaged until they elicit a response. Only in retrospect is an event registered as being "conscious." In that subtle sense, consciousness can be said to be an illusion. As Susan Blackmore[8] puts it, "We humans are unique because we alone are clever enough to be deluded into believing that there is a conscious 'I.'" (Open question: as the web nears the number of connections in a brain, should we expect it to become conscious?)

From this point of vantage, sentience appears, not as a phenomenon to be explained but as an epiphenomenon to be explained away. Most biologists appear to be quite content to do so, and get on with less squishy subjects. Philosophers are made of sterner stuff, foremost among them Thomas Nagel, who has wrestled with the mystery of consciousness for decades. His most recent book, *Mind and Cosmos*,[9] flings down a gauntlet to the entire materialistic worldview, which, Nagel argues, fails to deal with those two great unknowns—the origin of life and consciousness. In Nagel's view, neither is compatible with the scientific framework that we know and rely on; which leads him to assert, "The materialist neo-Darwinian conception of nature is almost certainly false." Nagel is thoroughly familiar with the literature, accepts that our conscious minds are products of evolution by natural selection, and has no covert inclination toward theism. But he is adamant that science cannot indefinitely postpone facing up to the elephant in the room: How can we merge our true knowledge of the external world with the equally valid knowledge of our interior selves into a coherent self-consistent worldview?

Nagel spells out the eventual goal:

> The essential character of such an undertaking would be to explain the appearance of life, consciousness, reason and knowledge neither as accidental side-effects of the physical laws of nature, nor as the result of intentional intervention in nature from without, but as an unsurprising if not inevitable consequence of the order that governs the natural world from within. That order would have to include physical law, but if life is not a physical phenomenon, the origin of life and mind will not be explained by physics and chemistry alone. An expanded, but still unified, form of explanation will be needed, and I suspect that it will have to include teleological elements.[10]

Heresy! I take Nagel seriously, at least up to a point, for his strictures mesh with my long-held suspicion that our understanding of life is not nearly bold enough to work with the weird universe discovered by physicists, one ruled by dark matter, dark energy, cosmic inflation, and quantum entanglement. The large-scale mapping of brains currently under way is sure to provide useful information, though I doubt that it will shine much light into the great mystery of mind.[11] But at the end of the day I am not ready to buy into a worldview that makes the world first and foremost an expression of a nebulous mind, practically inaccessible to our earthly brains, rather than solid matter. The conventional interpretation, which makes mind the output of ever more organized complex systems, may be a fallback position. But it has the merit of making a platform for concrete investigation, and I intend to hold onto it unless proven to be untenable. Hold it, but hold it lightly between two fingers, so that the wind can blow it away.

The Thinking Animal

To human eyes we have always been special, a long cut above other creatures. We have souls, the others don't. Presumably, if dogs could reflect and write they would extol "caninity" above all, but that is the point: dogs can't, and neither can other creatures. Even chimpanzees, our closest living relatives, fall well short of humans by many criteria, particularly mental ones. Objectively, humans are abundant around the globe while chimpanzees are rare, and it's the humans that run the zoo. That humans represent a higher level of functional organization than all other organisms seems to me self-evident, and so does the thesis that we owe our dominant place in today's world to our superior mental powers. As Darwin already noted in 1871, human minds differ from those of animals in degree rather than in kind. But the difference is glaring enough to mark a new era, the Anthropocene, and it cries out for explanation: Just what makes us special, and how did that come about?

The origin of humans and our place in the living world are emotionally charged issues that stoke the long-running conflict between science and religion. Polls indicate that as many as half the American public take humans to be products of divine creation, and reject the arguments for evolution; so much for respect for evidence. To those who prefer rationality it is obvious that humans are animals, mammals and specifically primates, with whom we share countless features of our anatomy, physiology, and biochemistry. Nor is there any doubt of the evolutionary origin of mankind: an extensive (albeit fragmentary) fossil record documents the transformation of the hominin line over the 6 million years since it diverged from that which gave rise to chimpanzees (Hominins refers to humans alone, hominids to all the African apes including humans). Chimpanzees and humans are so closely related that we share more than 98% of our gene sequences!

That is a startling number, for while the similarities between us and them are sufficient to make a visit to the chimpanzee enclosure somewhat uncomfortable, there is also no question that we differ physically as well as mentally. Something drastic happened over those 6 million years to produce a new kind of animal that habitually walks upright, has lost almost all its body hair, puts its hands to a thousand delicate tasks, talks incessantly about matters trivial and sublime, and features a large convoluted brain to keep control of its life.[12] The transformation of hominid skulls over time, illustrated in Figure 11.1, underscores the physical side of human evolution and raises the question, if we and chimpanzees are so much alike in our gene complement, what makes us so different?[13]

Begin with the genome. A difference of 2% out of some 3 billion base pairs leaves room for millions of sequence differences, and many molecular scientists look to those to account for the gulf between us and our cousins. Most of the critical changes probably involve not coding genes but elements of the regulatory web that determine when and where the protein products are deployed. The differences that underlie the disparities in skull evolution are presumably hard-wired, and many may be of this regulatory nature. But I suspect that our preoccupation with genes is blinding us to higher levels of order that are so conspicuous a feature of the human condition. One of the awkward discoveries to emerge from genomics is how few genes are required to make a human. Current estimates run to 21,000 coding genes, not all that many more than the 14,000 that specify a fruit fly (even *Escherichia coli* holds about 4,500 genes). To be sure, combinatorial arithmetic generates enormous complexity from the regulatory web, but even so it may turn out that many of the conspicuous differences between humans and apes may not be spelled out in the genes at all. There will be space for epigenetic differences, and also for developmental ones that arise from the way gene-specified elements are expressed within an inherited structural framework of cells and bodies, self-organized brains, and social organization. Of this we as yet know little, leaving the gulf between us and them a problem that has no satisfying answer.

The course of human evolution has been charted, at least in outline, anchored in a growing number of fossils from around the world. All humans living today belong to a single species, *Homo sapiens*; we know this because all of us can interbreed and produce fertile offspring. We are the sole survivor of a much larger family that featured fifteen to twenty members (Figure 11.1), which came and went extinct over the past 6 million years. What drove the

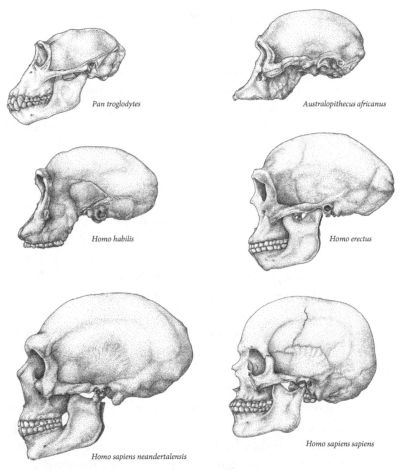

Pan troglodytes

Australopithecus africanus

Homo habilis

Homo erectus

Homo sapiens neandertalensis

Homo sapiens sapiens

Figure 11.1 Transformation of the hominid skull, drawn to scale. *Pan*, chimpanzee; *Australopithecus* ("Lucy"), *Homo habilis*; and *Homo erectus* are fossil hominins, dated to around 2.5, 2.0, and 1.0 billion years ago, respectively. *Homo neanderthalensis* and our own species, *Homo sapiens*, first appear 200,000 to 300,000 years ago. Courtesy of Deborah Maizels. Artwork copyright Debbie Maizels, Zoobotanica Scientific Illustration, 1994.

divergence of the human line from the other hominids was probably climate change: Africa began to dry up about 2.5 million years ago, causing forests to shrink while grasslands spread. This lured a population of forest-dwelling apes out of the trees and onto the savannah. The new habitat put a selective premium on upright posture, and on running over knuckle-walking: game animals may out-sprint the hunter, but are likely to be outrun in the end.

Walking freed the hands for other tasks, especially making tools; many animals including chimpanzees use simple tools, but only humans learned to knap flint. Open spaces also rewarded group living, which in turn favored social skills, communication, and intelligence. If we have this right, bipedal walking came first, then deft hands, the large human brain later still.[14]

By the time *Homo erectus* (the "upright man") appeared, about 1.8 million years ago, hominins were spreading around the globe and beginning to look distinctly like us (Figure 11.1). It is likely, though by no means certain, that *H. erectus* first evolved in Africa and radiated outward from there. The type specimens come from Indonesia, but putative relatives are known from China, Georgia, Africa, and as far west as Britain, but not from Australia or the Americas. *H. erectus* is associated with a distinctive stone object, the hand-axe, which marks the onset of the Acheulian period of cultural evolution and remained substantially unchanged for more than a million years thereafter. Oddly, though there must be hundreds of thousands of hand-axes scattered around museums and curiosity cabinets, no one knows for sure how and for what purpose these iconic objects were used. Tools? Probably, but they may also have served as ceremonial objects or as status symbols, not unlike the neckties that gentlemen used to wear.

As recently as 40,000 years ago, 4 or 5 hominin species shared the same space, all but one ending in extinction. To an observer at the time, it would not have been obvious which one was destined to make history; events—contingency, chance—probably played a large role in picking the winner, at least early on. The origin of our own species, *H. sapiens*, has occasioned much debate which is not altogether settled. The evidence comes from fossilized bones and from stone industries, but increasingly draws on sequences of ancient DNA. It sounds scarcely credible but is solidly established that informative DNA sequences can often be recovered from fossil bones. Of particular value has been mitochondrial DNA, which evolves quickly enough to track comparatively recent mutations, and which is inherited exclusively through the female line.

The prevailing wisdom holds that the crucial speciation event occurred in East Africa some 200,000 years ago, eventually producing a novel species endowed with some advantageous qualities. Just what these were is debatable, but a good case can be made for sharper and more effective stone tools, and for more advanced social skills including language and cooperation. In any event, modern humans apparently swept out of Africa and around the accessible globe; by 30,000 years ago modern humans were the dominant form of the species. Prior inhabitants, including our near-relatives the Neanderthals (Europe), Denisovans (Northeast Asia) and the "hobbits" of Flores in Indonesia, all became extinct, with relict populations clinging on as late as 10,000 years ago. Whether we had a hand in their demise is not certain, but it is clear that Cro-magnons with their sophisticated stone tools, bone sculptures, and stunning cave paintings were our immediate forebears. Were you to meet one on the street, dressed in contemporary fashion, you would not think him or her out of place.

Language, as Christine Kenneally points out in her recent article,[15] is the most distinctive biological trait on the planet, uniquely human; but there seems to be nothing in the human physique that accounts for it. Language is not encoded in the genome, it does not appear to be a singular adaptation, and it is not the inevitable outcome of our superior minds. Language, it seems, grew out of a platform of miscellaneous abilities, some ancient and shared with other animals, others more recent and specifically human. It is one of a constellation of cultural traits that evolved together, as much a cause of the expanding human mind as a consequence thereof.

There is no reason to believe that, having produced our own species, evolution is through with us. On the contrary, just since the invention of herding and then farming some 10,000 years ago there have been noticeable changes in human dentition. The capacity to tolerate lactose (milk sugar) has spread to become the norm among adults as well as babies, and patterns of disease and immune response have altered too. Will a culture that increasingly rests on communication (writing, printing, and now electronic) call forth heritable adaptations of its own? Such speculations are entertaining, but seem frivolous in the face of mounting threats to our very existence that stem from population growth, environmental deterioration, and uncontrollable technology. The outlook for the near future, say the next century, is very cloudy; and it is by no means certain that humankind will find the wisdom and the will to make drastic, permanent changes in time to avert catastrophe. Life has endured many mass extinctions over the past half-billion years, and will presumably survive no matter what we get up to. Indeed, I would not expect the imminent extinction of the human line: A species clever and resilient enough to make over the planet for its own benefit likely has the resources to endure and rebound. The question is whether the advanced, technological societies that we take for granted can make it through the eye of the needle.

A Cosmos Pulsing with Life?

There is an unforgettable scene in the first of the Star Wars movies, set in a sleazy bar in some outpost of the galaxy, where the multifarious denizens of the Galactic Empire mingle and chaffer. Up on stage a band of sky-blue elephants plays music—reads music too, no doubt. Does this amiable fantasy bear any resemblance to the real world? Is life, even intelligent life, a regular feature of the universe, or are we all alone?

Exobiology, the study of life beyond earth, is big business. A new Mars rover is even now drilling beneath the surface, seeking molecular indicators of life. Fly-bys sniff the atmospheres of Europa and Enceladus, moons of Jupiter and Saturn respectively, which may just conceivably host life in watery oceans beneath a thick crust of ice. Radio antennas scan the sky for signals from deep space. And every so often a supermarket tabloid regales us with a salacious tale of nubile young humans abducted by aliens for unspeakable purposes. With all the hype and hoopla it is important to state unequivocally that, as of the time of writing, there is not a shred of evidence that life exists beyond the confines of earth. None.

Absence of evidence has never inhibited speculation and reflection, nor should it. The question whether we share the cosmos with other kinds of life is one point of contact with ultimate concerns, and is rightly of passionate interest to thoughtful persons of all stripes. Moreover, where our solar system is concerned it is really not unreasonable to expect that, if life of some sort exists somewhere (or did so in the past), solid evidence will be forthcoming within the next few decades. Should that happen, the conversation will quickly shift to the nature of those creatures and their relationship to terrestrial life, and become part of normal science.

The void beyond the solar system is quite another matter, if only because of the incomprehensible vastness of space. Sixty years ago the cosmologist Frank Drake, one of the pioneers in the search for extraterrestrial intelligence, injected some cool reason into the cacophony of gab by estimating how many technologically advanced civilizations might plausibly exist within our own galaxy. Drake factored in such ponderables as the rate of star formation, the fraction generally similar to the sun, the number of planets and the fraction that hosts life, the fraction that attains a level of technology sufficiently advanced to be detectable, and their likely longevity. Numbers are necessarily wobbly, but Drake's own guesstimate puts the number of advanced civilizations at some 10,000 just in our own galaxy.

The odds in favor of life in the universe have continued to improve. The seemingly hopeless quest for exoplanets orbiting distant stars has proven spectacularly successful: some 4,000 are known at present, making planets a common feature of the cosmic landscape. Quite a few are situated in the "habitable zone," where liquid water can be expected to exist (water is known to be present on Mars, at least in the form of ice, and may have been abundant in the past). On the premise, unproven but almost universally shared, that life will arise wherever conditions allow, there will be many locations where life can exist; the question is, how likely is its emergence? Intelligent life is even more debatable. For what it's worth, my own opinion (shared with Conway Morris and others) is that life is uncommon in the universe but will be abundant locally wherever it did arise. Moreover, given that mind is itself a quality rewarded by natural selection (a good move in design space!) I would expect a significant fraction of the living planets to have brought forth intelligent life. These aliens might even look a lot like us: upright posture, walking legs, deft hands, eyes in front, and brains to keep it all together, simply because these too are good moves in design space. Sadly, unless the warp drive becomes reality we will never meet, because we are just too far apart. Even the closest known exoplanet, orbiting one of the nearest stars, is still more than four light-years away—over 24 trillion miles.

Can we ever know whether there is life beyond our solar system? I think we can, and within the lifetime of some readers of this book. Spectrographic telescopes are advancing rapidly, and should make it possible to gain information about the composition of some exoplanets' atmosphere. We know that the earth's atmosphere is grossly out of chemical equilibrium, featuring both methane, a strong reductant, and oxygen, a powerful oxidant. This is possible only because both are maintained in a steady state by biological activities: methane is produced by methanogenic Archaea, oxygen by cyanobacteria, algae, and plants. Should a powerful spectrograph detect oxygen in some planet's atmosphere, that body would instantly become a prime candidate to host photosynthesis; and then it would not be long before space tourists start to line up for tickets.

Epilogue

Comprehensible, but Complex and Perplexing

The mind for truth
Begins, like a stream, shallow
At first, but then
Adds more and more depth
While gaining greater clarity.

—Saigyō[1]

Biology is a broad church that has room for an ever-increasing range of specialties, both basic and applied. What unifies the subject is the cluster of questions surrounding life as a natural phenomenon: What is life, how do living things work, how are they related to one another and to nonliving matter, and how did they come to be as we find them? Perhaps only those who participated in the life of science during the outburst of creativity that followed the Second World War can fully appreciate how much we have learned, and how profoundly it has sharpened our perception of life. The discovery of how living things work is one of the signal accomplishments of the 20th century, and it is rather saddening to realize how little the general public knows of it, and how little it cares.

So where does the science of life now stand? Is it fully consilient with other natural sciences, or does it remain apart with its historic whiff of transcendence, something that can never be fully accounted for by mere chemistry and physics? Scientists by and large are doers, not thinkers. The great majority seem entirely content with the materialistic and reductionist attitude that now dominates scientific discourse, and supplies a solid platform for all manner of practical activities. To be sure, we all recognize that our knowledge is riddled with gaps; but we expect those to close over time in a seamless manner. Having been immersed in the work of discovery for half a century, I share much of this attitude, and this book can be read as a celebration of the hard-headed, mechanistic biology of our time. But this is also an occasion to stand back, take stock, and consider to what extent we can now account for the existence of even so mundane an object as a blade of grass (Box E.1).

The science of biology graduated from the 20th century with a wealth of factual knowledge and a clear sense of "what life is," with genes at the center of our worldview. By this account, living things are ensembles of interacting molecules that obey all the laws of chemistry and physics but are found in nature only in the context of life. The key to the phenomenon of life lies in the triad of information-carrying molecules: proteins, RNA, and DNA. In essence (and Renato Dulbecco's phrasing), life is the execution of the instructions spelled in the genes, which are carried on long molecules of DNA (RNA in some viruses). The genomic manual specifies all that living things are and do, including their manifold forms and functions. Evolution meshes nicely with that framework. Accidental errors in the transmission of genetic information are winnowed by natural selection for superior performance. All living things are related, but diverged as a result of variation, natural selection, and adaptation. All organisms, humans included, are products of a mindless, undirected process: the interplay of random variation (the genetic lottery) with selection by external circumstances. In essence, this is the position so forcefully articulated in 1970 by Jacques Monod in his masterful opus *Chance and Necessity*.[2]

Bleak as it may seem from a human perspective, it provides a comprehensible and straightforward account of how living things work and how they came to be; and in the past half-century it has become the conventional wisdom.

The gene-centered perspective on the ancient riddle "What is life?" is solidly grounded in empirical facts, but with the passage of time scientists have come to doubt that this is the only, or even the most compelling, interpretation of those facts. I am one of a minority, still small but growing, who suspect that the molecular revolution that transformed biology in the 20th century has just about run its course. Having ceased to be a wellspring of new ideas, it is degenerating into what the late Carl Woese (himself a molecular biologist par excellence) caustically dismissed as an "engineering discipline."[3] It turned out that the molecular viewpoint offers very little

Box E.1 What We Know and What We Don't

We understand quite well:

1. What life is.
2. How living things are constructed, from molecules to cells and beyond.
3. How living things make themselves from matter and energy extracted from the environment.
4. How heredity works.
5. That all living things are members of a single family, united by common descent, that has proliferated into a gigantic interactive web.
6. How that web has expanded over 4 billion years to colonize every habitable nook and cranny.
7. That expansion of the web generated mounting levels of complexity, autonomy, and agency, from bacteria to humans.
8. That the progression of life is underpinned by the interplay of heredity, variation, natural selection, and adaptation.
9. That there is no indication of direction by any external agency.

Much remains to be clarified, including:

1. How cells and organisms shape themselves.
2. How genes, self-organization, and physical structure collaborate.
3. How evolutionary innovations arise.
4. The pathways for the inheritance of acquired characteristics.
5. The early stages of cell and virus evolution.
6. How eukaryotic cells came into existence, and why they alone gave rise to higher organisms.
7. Whether some inherent evolutionary force favors the rise of autonomy, agency and mind.

Matters to Ponder
1. How did life begin?
2. What is mind (especially consciousness), and how did it come into being?

3. How could form, function, intention, morality, and reason emerge from mindless physics and chemistry?
4. Does life, especially intelligent life, exist elsewhere in the universe?
5. Does life have meaning, beyond its own perpetuation?
6. Down by the creek, a bumblebee assiduously samples the buffet of wildflowers. For all that we know about biochemistry and physiology, genes and evolution, the whole bee—buzz, bumble, and all—continues to elude our comprehension.

insight into some of the most prominent features of living things: organization, purpose, goal-oriented behavior, autonomy, integrity, even evolution. Progress in the quest to understand life, I hold, lies with an attitude that recognizes every cell and organism as a complex interactive system, and that ponders the nature and origin of biological organization. The discoveries of the molecular era stand, and are not about to be tossed overboard. On the contrary, they make up the platform on which a biology of complex systems is even now being built. But we are coming to see that reduction of organisms to the level of their molecules has been but a stage in the continuing search.

The holistic attitude that I favor has quite a long pedigree, reaching back to the German philosopher Immanuel Kant (1724–1804) and featuring a roster of contemporary contributors including Paul Weiss, Ludwig von Bertalanffy, Tracy Sonneborn, Lynn Margulis, Brian Goodwin, and Denis Noble.[4] The idea of biological systems is neither radical nor novel. Most biologists are well aware that living things make up a special category of objects distinguished not only by their extreme degree of orderliness (in the sense of regularity and predictability) but also by the sort of purposeful order designated organization. Only in living things do myriads of special molecules, obeying only the laws of chemistry and physics, come together into integrated structures that display functions, forms, behavior directed toward persistence and reproduction, and the capacity for evolution by variation and natural selection. The molecular revolution, with its tight focus on the molecular parts, relegated the organismal whole to the "soft" margins of our science; systems biology restores the organism to its rightful place at the core of biology. The molecular parts are indispensable for an understanding of the organism, but definitely not sufficient: physiology, behavior, and evolution are emergent properties that only become apparent at the level of systems. These are now returning to center stage, within a conceptual framework that underscores the web rather than its individual strands.

From this perspective the genetic machinery appears as the cardinal feature of a greater entity, the cell, which is the proper unit of life. Life, we might say, is what cells do—not solely what their genes prescribe. Cells are multimolecular systems, intricately and purposefully organized at all levels from molecules and genes to physiology, functions, and durable structures. They draw matter and energy into themselves, maintain their identity, and reproduce their own kind. To these ends energy generation, compartments, membranes, and a cytoskeleton are just as essential as genes; and they, together with the organized cell as a whole, constitute what is reproduced with every cell cycle. Here we encounter another set of commonalities: membranes, lipid bilayers, ATPases, and energy coupling by proton currents, found (with variations) in all cells. Moreover, while genes specify molecular structure (sometimes directly, more commonly indirectly), they specify neither spatial architecture nor the higher levels of order: genes replicate, cells reproduce. As we track the gyre of complexity from the level of single cells to that of multicellular organisms and beyond, the distance between what is spelled in the genes and what is displayed by the body grows ever longer. Do genes determine, say, human features? Yes, and sometimes quite directly; a person who carries the mutation responsible for Huntington's disease is all but certain to develop that condition. But such a straightforward relationship between genotype and phenotype is uncommon. More often, the phenotype is the outcome of a complex interplay between multiple genes, filtered through multiple layers of organismic structure. My assertive nose has run in my mother's family for generations, but there is no one gene for "nose." The nose's shape is the output of an interactive system composed of gene-specified elements.

I take it for granted that in order to appreciate the integrity, the wholeness of organisms, it is essential to adopt the systems perspective; genes are part of that web, and they hold a special place in it. The genome is neither the blueprint for a cell nor the fountainhead of a linear chain of command; its role is to provide a vital and relatively permanent database, a registry of what is possible for any particular creature. Living organisms are the only known systems whose parts are specified by the system itself. The triad of informational molecules—DNA, RNA, and proteins—specifies the chemical underpinnings of all cell operations and assures their accurate transmission from one generation to the next. Critically, information (specifically, sequence information) flows unidirectionally from the level of nucleic acids to that of protein, never the reverse. The genome is generally a fixed feature, not subject to revision by cellular mechanisms; organisms cannot rewrite their genetic manual according to need. Changes do occur (by mutation, lateral gene transfer, symbiosis, or whatever), and cells possess mechanisms to promote and control such happenings. But the resulting modifications to the genome serve to benefit the population, not the individual; and they are mostly random events, not specifically targeted to the needs of the moment. It is quite extraordinary, and surely tells us something fundamental, that the molecules and procedures involved in the transmission of genetic information are universal, found in all living organisms with only minor variations. They bespeak a singular pattern, invented at the very dawn of life, and jealously guarded and preserved for more than three billion years.

The systems perspective represents a notable expansion of our vision, which resonates further when we turn from physiology (how living things work) to evolution (how they got to be that way). The traditional focus on gene mutations has been enlarged by a panoply of mechanisms that bring in genes from abroad (lateral gene transfer, viruses, symbiosis), as well as adaptations that tweak the genetic instructions. The inheritance of acquired characteristics, changes to the workings of an organism rather than to its genes, is no longer taboo: they can become naturalized and may be favored by natural selection. Symbiosis and lateral gene transfer are well-established examples. But it is important to emphasize that the Darwinian framework is being expanded, not replaced. As I read the evidence, it appears that only changes that have been incorporated into the genome are sufficiently durable to bring forth new species. The bottom line remains unchanged: living things are complex multimolecular systems, built of gene-specified elements, that make themselves from matter and energy extracted from the environment, reproduce their own kind, and are shaped over time by heredity, variation, and natural selection.

Let us now look once more at life's history. It is undeniable that the overarching pattern is one of increase and expansion, from sparse and small single cells that look just like today's Bacteria and Archaea to overwhelming profusion, multicellular creatures of great size, higher functional organization, and operational capacities culminating (so far) in the human mind. We know of only two processes that generate organization: intelligent design, as in engineering, and evolution by heredity, variation, and natural selection. This puts the spotlight on a fraught question: do random variation and natural selection fully account for the burgeoning of purposeful order, homeostasis, goal-oriented behavior, and minds that contemplate their own nature; or is something more required to lend direction to the game of chance and necessity?

Most scientists today are persuaded that Darwin's ratchet is sufficient, and anyone who postulates direction or intent is pandering to religious mystification (Monod would have been disgusted by the "spinelessness" of all such efforts). Indeed, as far as the evidence goes, evolution has no intrinsic direction, no design and no purpose. It is the outcome of blind chance, channeled by natural selection for reproductive advantage; and that's the way it is, whether we humans are comfortable with it or not. The public, and more than a few scientists, remain skeptical. For example, in an engaging and provocative recent book titled *Purpose and Desire: What Makes Something "Alive" and Why Modern Darwinism Has Failed to Explain It*,[5] Scott Turner argues the case to the contrary: in his view purpose, intention, and direction are primary features of life. They are not products of natural selection but antecedent to it, inherent to the fabric of our world. There is, to the best of my knowledge, no positive evidence whatsoever that purpose, desire, or intention exert any influence on the course of evolution; and it is tempting to dismiss Turner's musings as lacking in substance. In fact, many of his concerns are valid enough: "Darwinism" supplies a useful and comprehensive framework, but leaves room for plenty of uncertainty. The open questions need to be put on the table, not swept out of sight behind a curtain.

My own unease with the prevalent view of life centers on the nature, generation, and origin of biological organization. Remember that "organization" is defined as "order that has purpose," and living things flaunt their purposeful nature in every aspect of their being. The physical structure of cells, their metabolism, self-regulation, and behavior—even many facets of mind—are obviously directed to one purpose: to persist and reproduce. If there is any higher purpose it does not leap to the eye. What we have learned of evolution leaves no doubt that the undirected operations of heredity, variation, natural selection, and adaptation can bring forth ever higher levels of organization, and have done so. What I remain unsure of is whether living organization is wholly the product of this mechanical iteration, or draws on some deeper well.

Take the form and architecture of cells, the most basic level of the phenomenon of life. How do cells build the next generation of their own kind? To make sense of that we had to invoke, in addition to genomic information, the abstruse physical principle of self-organization. The term refers to the observation that, in systems kept off equilibrium by the flow of energy, order quite commonly emerges from the interactions of elements (molecules, for example) that obey only local rules and have no idea that they are part of a larger system. The term surfaces at every level of biology, from cell division and embryonic development to the coordinated movements of a school of fish. It's hard to grasp but quite real, and seems to be baked right into the constitution of the universe. There seems to be some "force" that drives the emergence of layer upon layer of cosmic order: subatomic particles, atoms and molecules, dust, stars and planets, galaxies, and beyond. Variation and selection feature in one of these layers, but self-organization looks like something far deeper.

The fog of uncertainty lies thickest over the origin of life which, after eighty years of reflection and experimentation, holds fast to its secrets. Evolution by heredity, variation, natural selection, and adaptation explains to the satisfaction of most contemporary scientists how and why life prospered, diversified, and climbed up "Mount Improbable" (Richard Dawkins).

But Darwin's ratchet can only take hold once a basic level of functional organization has been achieved. It does not readily explain the genesis of organized structures in the first place, and that has emerged in recent years as biology's most stubborn riddle. Moreover, the fact that all living things on earth share a common ancestry implies that, whatever happened, it was in some sense a unique event. A singular creation? Not necessarily; it may be, for example, that all contemporary living things descend from the handful of survivors of some primordial cataclysm. Whatever the explanation, seems that at some early stage in evolution life had to squeeze through a tight bottleneck.

I keep wrestling with the ultimate origin of organization without reaching a satisfying conclusion. Where did those (hypothetical) protocells come from, endowed with a rudimentary level of structural and functional order that natural selection could get its teeth into? Is natural selection from the outset, perhaps for persistence and stability, the correct key? Self-organization, especially the elusive capacity of systems kept far from equilibrium by energy flow to generate order on a scale far above the molecular, probably enters here. It seems implausible that self-organization by itself can bring forth agency and autonomy from the meaningless jostling of molecules. More likely, self-organization works together with natural selection to bridge the gap between the molecular scale and the cellular. But when all has been mulled and chewed, I cannot get away from those universal necessities of life that must have come together just to get evolution

started: genes, energy, metabolism, and physical structure, at least in rudimentary form. All genes, as far as I know, come from preexisting genetic
sequences. Membranes never arise de novo, only from previous membranes.
Nothing much happens without energy input. And it always takes a cell to
make a cell. Genes replicate, cells reproduce, organisms evolve. It seems that
a significant level of organization had to be present from the beginning; and
yet, unless we posit creation by some transcendent mind and will, it must all
have originated from unorganized chemicals, presumably on a hostile earth.
Nothing would be so reassuring as a credible model system! Is there some alternative that has eluded all our speculations?

Loren Eiseley (1907–1977), a naturalist, philosopher, and poet whose
writings played a formative role in my personal evolution, struggled with the
deep questions all his life. In his autobiography he summed up his conclusions
in lines that continue to resound in my head. "In the world there is nothing to
explain the world. Nothing to explain the necessity of life, nothing to explain
the hunger of the elements to become life, nothing to explain why the stolid
realm of rock and soil and mineral should diversify itself into beauty, terror
and uncertainty."[6] Writing now in the twilight of my own journey but forty
years of spectacular progress later, Eiseley's elegiac ruminations still have the
ring of truth.

Glossary

Adaptation In evolutionary biology, any change in the structure or functions of an organism that make it better suited to its environment.

Adenosine diphosphate (ADP) See adenosine triphosphate.

Adenosine triphosphate (ATP) The universal energy currency of living things. The molecule consists of adenine, ribose, and three phosphoryl groups; release of the terminal phosphoryl group gives rise to ADP, and can be coupled to the performance of work.

Algae In this book, used generically to designate photosynthetic protists.

Allele One of two or more alternate states of a gene that typically arise by mutation.

Amino acids Small, water-soluble compounds that possess both a carboxyl group (-COOH) and an amino group ($-NH_2$). Proteins are polymers made up of a characteristic set of twenty amino acids.

Archaea Microbial clade originally defined by ribosomal RNA sequences and associated with extreme environments, now known to be widespread. Archaea constitute one of the three domains of life.

Archaeon An organism classified in the domain Archaea.

ATP See *adenosine triphosphate*.

ATP synthase A class of complex enzymes pivotal to energy conversion, that links ATP chemistry in the cytoplasm to the current of protons across the plasma membrane. In bacteria, mitochondria, and plastids, ATP synthase catalyzes the production of ATP during oxidative and photosynthetic phosphorylation.

Autotroph An organism that can derive all its carbon from CO_2.

Bacteria Diverse microbial domain that includes most of the familiar lineages, as well as the ancestors of both mitochondria and plastids. When spelled with a lowercase *b*, refers to the traditional informal term for prokaryotes.

Bacteriophage ("phage") A virus that is parasitic on bacteria.

Blastula The first stage in the development of an animal embryo, consisting of a hollow ball of dividing cells.

Cambrian explosion The abrupt appearance of animal phyla at the beginning of the Cambrian era.

Catalyst A substance that increases the rate of a chemical reaction without itself undergoing permanent chemical change. Enzymes are the chief catalysts of biochemical reactions.

Cell The structural and functional unit of living organisms. Cells vary in size, but most are microscopic. Many organisms consist of but a single cell (bacteria and protists), others (plants, fungi ,and animals) are multicellular.

Cell wall The strong and relatively rigid envelope of many cells, external to the plasma membrane. Cells of plants, fungi, protists, and bacteria are mostly walled; animal cells are not.

Central dogma The proposition that sequences of nucleotides in DNA and RNA can specify the amino acid sequences of proteins, but the reverse is never true.

Chloroplast A membrane-bound organelle that is the locus of photosynthesis in plants. See also *plastid.*

Chromosome A thread-like structure found in cell nuclei, which carries genes in linear array.

Ciliate A protozoan whose surface is studded with cilia, which serve as the organs of motility.

Clade A group of organisms comprising a common ancestor and all its descendants.

Complexity The condition of a system made up of many interacting parts, whose behavior is not easily deduced from the properties of those components.

Cyanobacteria A phylum of Bacteria characterized by the possession of photosynthesis of the same kind as that of plastids.

Cytoplasm The jelly-like material that fills cellular space, exclusive of the nucleus and other organelles.

Cytoskeleton A network of microscopic filaments and fibers that pervades the cytoplasm of both eukaryotic and prokaryotic cells. Its functions include motility, cell division, intracellular transport, and secretion.

Deoxyribonucleic acid See *DNA.*

DNA (Deoxyribonucleic acid) A nucleic acid composed of two intertwined polynucleotide chains; the sugar is deoxyribose. DNA is the genetic material of all cells and many viruses.

Domain The highest level of biological classification. There are three domains—Archaea, Bacteria, and Eukarya.

Ediacaran fauna A large and diverse group of multicellular organisms that flourished about 600 million years ago. Named after the Australian station where the fossils were first found.

Emergence The appearance of new properties in a system that were not present nor easily predictable from the properties of the components.

Endosymbiosis A symbiosis in which one of the partners (the endosymbiont) resides within the cytoplasm of its host.

Energy The capacity to do work. In biology, the capacity to drive processes that do not take place spontaneously, such as the synthesis of complex molecules.

Entropy A portion of the energy of a system that is unavailable to do work. In a wider sense, a measure of disorder; as disorder increases, so does the entropy.

Enzyme A protein that acts as a catalyst in biochemical reactions.

Epigenesis The approximately stepwise process by which genetic information, modified by environmental factors, is translated into the substance and functioning of an organism.

Epigenetic inheritance Refers to features that are heritable but not encoded in the nucleotide sequence of genes.

Escherichia coli (E. coli) A normal inhabitant of the human gut, classified in the phylum Proteobacteria. Of all living organisms, the one most fully understood.

Eukarya (Eucarya) The domain of life that contains all the eukaryotes, both unicellular and multicellular.

Eukaryotes (Eucaryotes) Organisms whose genetic material is enclosed in a true, membrane-bound nucleus. Eukaryotic cells typically also have a cytoskeleton, internal membranes, and organelles.

Flagellate A motile protist equipped with flagella.

Flagellum, Cilium A relatively long, whip-like structure present on the surface of many cells that serves as an organ of motility. The flagella of Bacteria, Archaea, and Eukarya are each entirely different in structure and operation.

Function In biology, a substance or processes has a function if it contributes directly to the production, maintenance, or reproduction of an organism. In this book, "purpose" is sometimes used as an informal synonym for function.

Gastrula The stage of embryonic development that succeeds the blastula. Produced by invagination of the blastula, generating the embryo's germ layers.

Genetic code The table of correspondences that specifies the particular triplet of nucleotides in DNA or RNA that specifies each amino acid in a protein.

Genome The total gene complement of an organism.

Holism The doctrine that a biological system is more than the sum of its components. To understand the organism it is necessary to know its parts, but that dissection always leaves an unexplained residue that turns on the organization of those parts.

Hominid The evolutionary lineage that includes humans as well as the great apes. *Hominin* refers to species more closely related to humans than chimpanzees, our nearest relative among the apes.

Homology Similarity of structure or function due to common ancestry.

Horizontal gene transfer See *Lateral gene transfer.*

Hypha (plural Hyphae) The characteristic form of a growing fungus: a filament of cells that extends at the tip.

Information An abstract and slippery concept that refers to useful knowledge an organism has about its environment or its internal state. In biology, information is a kind of power: energy is the power to do, information is the power to direct what is done (Francois Jacob).

Ion current The circulation of ions, usually protons, across a membrane. A key element in biological energy conversions.

Kingdom In taxonomy, a high level of classification. Traditional kingdoms include animals, fungi, and plants.

Last universal common ancestor (LUCA). A hypothetical organism ancestral to all three domains of life.

Lateral gene transfer Transfer of genes by mechanisms other than cell reproduction.

LECA (Last eukaryotic common ancestor) The hypothetical organism most recently ancestral to all extant eukaryotes.

Lipid Generic term for a diverse group of organic substances that are soluble in organic solvents such as benzene and chloroform, but not in water. Examples include fats, phospholipids, and steroids.

Membranes In this book, the thin sheets made of lipids and proteins that form the boundaries of cells, organelles, and intracellular compartments.

Meristem A plant tissue consisting of actively dividing cells that give rise to organs such as roots and shoots.

Messenger RNA The product of transcription of protein-coding genes, whose function is to carry sequence information from genes to ribosomes.

Metabolism The sum of all the chemical reactions that take place in an organism. Compounds that take part in, or are formed by, these reactions are called *metabolites*. A sequence of reactions that generates (or degrades) a metabolite is called a *metabolic pathway*.

Metazoa Animals whose body is composed of many cells grouped into tissues and organized by a nervous system. The term includes all vertebrates and invertebrates, but excludes protists.

Methanogenesis A metabolic pathway by which CO_2 is reduced to methane, with the production of useful energy and cell constituents. All methanogens belong to the domain Archaea.

Mitochondria Organelles of eukaryotic cells that are the locus of respiration and ATP production. The powerhouses of eukaryotic cells.

Mitosis The stately dance by which eukaryotic cells divide into two daughter cells, each of which has a nucleus containing the same number and kind of chromosomes as the mother cell.

Modern synthesis The theory of evolution as reformulated in the middle of the 20th century.

Monomer See *Polymer.*

Morphogenesis The development of the form and structure of a cell or organism.

Mutant An organism or gene that has undergone a mutation.

Mutation A heritable change in the genetic material of a cell that may cause it and its descendants to differ from the normal in appearance or behavior.

Neuron Nerve cell. An elongated branched cell specialized for the conduction of electrical impulses.

Neutral evolution Evolutionary changes in organisms or populations that carry neither selective benefit nor penalty.

Nuclear membrane The membrane that delimits the nucleus and regulates the flow of material between nucleus and cytoplasm. By common understanding, eukaryotic nuclei are bounded by a membrane, prokaryotic ones are not.

Nucleic acid A large and complex macromolecule consisting of a chain of nucleotides. There are two types, DNA and RNA.

Nucleotides The building blocks of nucleic acids. Each nucleotide consists of a nitrogenous base, a sugar, and one or more phosphate groups.

Nucleus, nucleoid The large, membrane-bound body embedded in the cytoplasm of eukaryotic cells that contains the genetic material. A nucleoid is the corresponding structure in prokaryotes, which is not enclosed in a membrane.

Order A state in which the components are arranged in a regular, comprehensible, or predictable manner. Various kinds and degrees of order are represented by the letters of the alphabet, the arrangement of keys on a keyboard, wallpaper, and the sequence of amino acids in a protein.

Organelle A minute structure within a cell that has a particular function. Examples include the nucleus, mitochondria, and flagella.

Organism An individual living creature, either unicellular or multicellular.

Organization Purposeful or functional order, as in the arrangement of the parts of a bicycle or a skeleton.

Oxidative phosphorylation The enzymatic generation of ATP coupled to the transfer of electrons from a substrate to oxygen. This is the chief mechanism by which aerobic organisms produce ATP.

Phagocytosis The process by which cells engulf and digest minute food particles.

Phospholipids A group of lipids that contain both a phosphate group and one or more fatty acids. Major constituents of biological membranes.

Photosynthesis The chemical process by which green plants (and many other organisms) produce organic compounds by the use of light as an energy source.

Phylogenomics In evolution, large-scale reconstruction of a phylogeny using sequence data.

Phylogeny Evolutionary relationship among organisms or their parts (e.g., genes).

Phylum A category used in classification, especially of animals; examples include *Protozoa* or *Mollusca*. Botanists and microbiologists commonly prefer the term "Division."

Plasma membrane (also Cytoplasmic membrane) The membrane that forms the outer limit of a cell and regulates the flow of nutrients in and waste products out.

Plastid An organelle of photosynthetic organisms, generally specialized for photosynthesis. The plastids of plants are usually referred to as *chloroplasts*.

Polymer A molecule or complex of molecules composed of repeating elements (monomers). For example, proteins are polymers of amino acids, starch is a polymer of glucose.

Polynucleotide Any polymer of nucleotides, including DNA and RNA.

Polypeptide A peptide containing ten or more amino acids. Proteins are long polypeptides.

Prokaryotes (Procaryotes) Organisms whose genetic material is not separated from the cytoplasm by a nuclear membrane, and which generate energy at the plasma membrane. More generally, a grade of cellular organization lacking a true nucleus and most organelles. Both Bacteria and Archaea come under this term.

Protein A large molecule consisting of one or more polypeptide chains, commonly hundreds or thousands of amino acids in length.

Proton A hydrogen atom that carries a positive charge due to loss of its electron.

Proton pump An enzyme that uses energy to translocate protons across a membrane, generally against a gradient of concentration or potential. Examples include ATP synthases and respiratory chains.

Protist A member of the kingdom *Protista (Protoctista)*, which includes the unicellular eukaryotes and some multicellular lineages.

Protocell In this book, a loose term to designate hypothetical precellular entities that may have lacked genes and proteins.

Protozoa In this book, used generically to designate nonphotosynthetic protists such as amoebas and ciliates.

Punctuated equilibrium A model of the pattern of evolution, which proposes that long periods of relative stasis are punctuated by episodes of rapid change.

Receptor A molecule on the cell surface whose function is to detect a particular stimulus and elicit a response.

Redox chain A cascade of enzymes that carry out successive steps of reduction and oxidation. Respiratory chains are the prime examples.

Reductionism The doctrine that the properties of a complex system can be largely, or even wholly, understood in terms of its parts or components.

Replication The production of an exact copy. Usually employed in reference to the replication of DNA, in which one strand supplies a template for the synthesis of a complementary strand, which is then copied once more to reproduce the original.

Respiration The utilization of oxygen. In cell biology, the oxidative degradation of organic substances with the production of metabolites and energy. Oxygen is the usual oxidant, but sulfate or nitrate sometimes take its place in "anaerobic respiration."

Respiratory chain The biochemical basis of respiration. A cascade of proteins and other molecules that carries electrons from substrate to oxygen.

Ribosomes Intracellular organelles that carry out protein synthesis, found in all cells. They are composed of two subunits, several species of RNA and as many as fifty proteins.

Ribozyme A biological catalyst composed of RNA (unlike enzymes, which are proteins).

RNA (Ribonucleic acid) A nucleic acid whose sugar is ribose. Examples include ribosomal RNA, transfer RNA, and messenger RNA.

RNA polymerase An enzyme that catalyzes the synthesis of RNA from its nucleotide precursors, typically using an existing strand of DNA or RNA as a template. The agent of gene transcription.

Second Law (of thermodynamics) A fundamental principle of physics which states that the spontaneous direction of events points downhill: hot bodies cool, order tends to dissipate, complex molecules fall apart, we ourselves age and die. To make some process go uphill (warm the coffee, maintain order, synthesize a protein) requires an input of energy and the performance of work. Even so, in the world at large (process plus the energy source), the Second Law always prevails.

Self-assembly A mode of self-organization in which supramolecular order emerges from the association of molecules without any input of external energy or information. Examples include the polymerization of tubulin into microtubules. Instances in which an external source of energy is required are referred to as self-construction; the mitotic spindle is a case in point.

Self-organization In this book, the emergence of supramolecular order without reference to any external template or plan, from the interaction of many molecules each of which obeys only local rules.

Sequence The order of amino acids in a protein, or of nucleotides in a nucleic acid. The determination of that order is described as *sequencing*.

Species In taxonomy, a group of similar and closely related individuals that can usually breed among themselves.

Stromatolite A rocky, cushion-like mass, produced by the growth of large populations of cyanobacteria and algae. They are commonly found fossilized, but living ones still exist in favorable locales.

Structural heredity The transmission of structural features by cellular continuity, without participation of genes. Also called *cell heredity.*

Symbiosis Living together; an interaction between individuals of different species that is beneficial to both partners.

System An entity composed of elements that interact, or are related to one another, in some definite manner. A bicycle and a cell are systems; a lump of granite is not.

Taxonomy The practice or science of classification.

Teleonomy The apparent purposefulness of living organisms, due to evolution by heredity, variation and natural selection. Contrast with teleology, the doctrine that living things are products of purposeful design.

Transcription Assembly of an RNA molecule complementary to a stretch of DNA. This is the first stage in protein synthesis, and represents the transfer of sequence information from DNA to RNA.

Translation Assembly of a protein by a ribosome, using messenger RNA to specify the sequence of the amino acids.

Vectorial Having a direction in space.

Vesicle A small, membrane-bound sac (usually filled with fluid) within the cytoplasm of a living cell.

Virus A particle too small to be seen with the light microscope or to be retained by filters, but capable of reproduction inside a living cell. Viruses have very limited metabolic capacities and are obligatory intracellular parasites.

Work A process that runs counter to the spontaneous direction of events, and therefore depends on an input of energy. For example, protein synthesis represents work, but they degrade spontaneously.

Notes

Preface

1. Midgley 1992.

Chapter 1

1. Monod 1971, 9.
2. Nearly one hundred definitions of life can be found in the literature. Two succinct ones that I value are: (1) Living things are autopoietic systems; they make themselves (F. Varela and U. Maturana); and (2) Living things are objects that possess whatever is required for evolution by heredity, variation, and natural selection (J. Maynard Smith). These definitions overlap but are not identical: Viruses would be alive according to Maynard Smith, but not according to Varela and Maturana.
3. Beck 1957.
4. Riedl 1978.
5. O'Malley 2014.

Chapter 2

1. Popper 1972, 84.
2. For an account of how the concept of prokaryotes and eukaryotes evolved, see Sapp 2005.
3. Carl Woese's monumental contributions are celebrated by Pace et al. 2012.

Chapter 3

1. Alberts 1998, 291.
2. Phillips and Milo 2009.
3. Bioenergetics is complicated. For introductions see Mitchell 1979; Harold 1986, 2001; Nicholls and Ferguson 1992.
4. Bioenergetics is complicated. For introductions, see Mitchell 1979; Harold 1986, 2001; Nicholls and Ferguson 1992.
5. Reading 2006.

Chapter 4

1. Woese 2004, 179–180.
2. Schroedinger 1944.
3. Riedl 1978.
4. Capra and Luisi 2014.
5. Queller and Strassman 2009.
6. Jacob 1973.
7. Dulbecco 1985.
8. Harold 2005.
9. Lartigue et al. 2007; Gibson et al. 2010; Hutchinson et al. 2016.
10. Harold 2005; Cavalier-Smith 2010.
11. Bromham 2016.
12. Doolittle 2013.
13. Shapiro 2011.
14. Bromham 2016.
15. Mayr 1982.
16. Capra and Luisi 2014.
17. Harold 2005.
18. Margolin 2009; Jiang et al. 2015; Surovtsev and Jacobs-Wagner 2018.
19. Harold 2005; Karsenti 2008; Rafelski and Marshall 2009.

Chapter 5

1. Albert Einstein, quoted in Schlipp 1949, 33.
2. There is a considerable literature making the case that evolutionary theory needs fundamental revision. For a sample see Goul, 1982; Goodwin 1994; Margulis and Sagan 2002; Shapiro 2011, 2013; and Noble 2014.
3. See Gould, 1982; Goodwin 1994; Margulis and Sagan 2002; Shapiro 2011, 2013; and Noble 2014.
4. Most recently by Behe 2019; See also a scathing review by Lents et al. 2019.

Chapter 6

1. Doolittle et al. 2003, 39.
2. Harold 2014; Koonin 2010.
3. Margulis and Schwartz 1998.
4. Woese et al. 1990; Pace 2009.
5. McInerney et al. 2008; Koonin 2015; Booth et al. 2016.
6. Arndt and Nisbet 2012; Allwood 2016.

7. Woese 1998; Becerra et al. 2007.
8. Fuchs 2011; Weiss et al. 2016.
9. Lyons et al. 2014.
10. Bruessow 2009; Krupovitch et al. 2011; Forterre et al. 2014.
11. Knoll et al. 2006.
12. McInerney et al. 2014; Eme et al. 2017; Lopez-Garcia and Moreira 2019; Imachi et al. 2020.
13. McInerney et al. 2014; Eme et al. 2017; Lopez-Garcia and Moreira 2019; Imachi et al. 2020.
14. Lane and Martin 2010.
15. Dacks et al. 2016.
16. Lane and Martin 2010; Lane et al. 2020.
17. Lane and Martin 2010; Lane et al. 2020.

Chapter 7

1. Pross 2013, 186.
2. Books that survey the origin of life include Fry 2000; Hazen 2005; Pross 2012; Harold 2014; Lane 2015; and Kauffman 2019.
3. Pross 2003, 2012, 2013; Pascal and Pross 2016; the quotation comes from Pross 2003.
4. Pross 2003, 2012, 2013; Pascal and Pross 2016; the quotation comes from Pross 2003.
5. Orgel 2004; recent contributions include Service 2019 and Muchowska et al. 2019.
6. Adamala and Szostak 2013; Pressman et al. 2015; and Gross 2016.
7. Lane and Martin 2012; Sousa et al. 2013; Russell et al. 2013; Deamer and Georgiou 2015.
8. Orgel 2004; Pressman et al. 2015; Higgs and Lehman 2015.
9. Scepanski and Joyce 2014.
10. Taylor 2016.
11. Ashkenasi et al. 2017.
12. Trevors and Abel 2004.

Chapter 8

1. Blum 1968, 194.
2. Knoll et al. 2006; Dacks et al. 2016; Betts et al. 2018.
3. Falkowski and Isozaki 2008; Lyons et al. 2014; Falkowski 2015; Fox 2016.
4. Falkowski and Isozaki 2008; Lyons et al. 2014; Falkowski 2015; Fox 2016.
5. Brasier 2009.
6. My understanding of evolution and the forces that drive it were shaped by the writings of John Tyler Bonner, Richard Dawkins, Stephen Jay Gould, Daniel Dennett, Ernst Mayr, and George Gaylord Simpson. Input for the present volume came from

Maynard-Smith and Szathmary 1995; McShea and Brandon 2010; Koonin 2011; Ward and Kirshvink 2015; and Falkowski 2015.

7. Coleman et al. 2017.
8. Wicken 1987; Schneider and Sagan 2005.
9. Wicken 1987.
10. Pedrós-Alió and Manrubia 2016.
11. Parfrey and Lahr 2013; Richter and King 2013; Sebé-Pedrós et al. 2017.
12. Parfrey and Lahr 2013; Richter and King 2013; Sebé-Pedrós et al. 2017.
13. Blount et al. 2012; McLysaght and Hurst 2016; Good et al. 2017.
14. McShea and Brandon 2010; Shapiro 2011; Brunet and Doolittle 2018.
15. Numerous new genes must have been created in the course of evolution; where did they come from? The majority probably arose from preexisting genes, by duplication followed by divergence driven by mutation. It now appears that some, perhaps many, stem from the noncoding ("junk") DNA that makes up a large fraction of the eukaryotic genome. And some, perhaps, arose by reverse-transcription of cellular RNA. Note that DNA is always copied from some preexisting nucleic acid template; it never comes into existence unaided, from its constituent nucleotides alone.
16. Bonduriansky and Day 2009; Jablonska and Raz 2009.
17. Advocates include Shapiro 2011 and Noble 2017.
18. Lynch 2007; McShea and Brandon 2010; Brunet and Doolittle 2018.
19. Gould and Eldredge 1993; Koonin 2011.
20. Szathmary 2015; O'Malley and Powell 2016.
21. Brasier 2009; Fox 2016.
22. Rosslenbroich 2006, 2014; McShea 2015; Kauffman 2019.

Chapter 9

1. Darwin1859, 459.
2. Data from Pedros-Alio and Manrubia 2016.
3. For a glimpse of the deep hot biosphere, see Jorgensen and D'Hondt 2006, and Jorgensen 2011.
4. Mayr 1997, 196.
5. Goodwin 1994.
6. Pradeu 2016. See also Pepper and Herron 2008; Queller and Strassman 2009.
7. Gilbert et al. 2012.
8. Gilbert et al. 2012.
9. Falkowski 2015. Also Falkowski et al. 2008.
10. Lovelock 2003; Ruse 2013.
11. Lenton and Latour 2018.
12. Doolittle 2019; Doolittle and Inkpen 2018.

Chapter 10

1. Mayr 1997, 152.
2. Davidson and Erwin 2006.
3. Gurdon and Bourillot 2001; Rogers and Schier 2011; Gilmour et al. 2017. For an invaluable brief introduction, see Wolpert 2011.
4. Mathur 2006; de Vries and Weijers 2017.
5. Harold 2002.
6. Riquelme 2013; Riquelme et al. 2018.
7. Hepler et al. 2013; Hepler 2017.
8. Davidson and Erwin 2006; Carroll 2008. See DiFrisco and Jaeger 2020 for an illuminating discussion of just how genes function as causal components in much grander systems, that may range in size from the molecular level to the organismal.
9. Carroll 2008.
10. Jacob 1973, 313.
11. Waddington 1957; Huang 2011; DiFrisco and Jaeger 2020.

Chapter 11

1. Chandler 2001, 2.
2. Shorto 2008.
3. Balus'ka and Reber 2019.
4. Baker et al. 2001; Bromham 2016.
5. Kandel 2001; Josselyn et al. 2015; Ramirez 2018.
6. Dennett 2018.
7. Dennett 2017, 2018; Koch 2018; Suddendorf 2018; Blackmore 2018.
8. Blackmore 2018.
9. Chandler 2001; Nagel 2012, 32.
10. Chandler 2001; Nagel 2012, 32.
11. Fregnac 2017.
12. Carroll 2005.
13. Carroll 2005.
14. Gee 2015; Laland 2018; Wong 2018.
15. Kenneally 2018.

Epilogue

1. Saigyo (1118–1190, Japanese priest and poet).
2. Monod 1970.

3. Woese 2004.

4. For some recent perspectives from different viewpoints, see Capra and Luisi 2014; Goodwin 1994; Harold 2001; and Noble 2006 and 2017.

5. Turner 2017.

6. Eiseley 1975, 242.

References

Adamala, K., and Szostak, J. 2013. Nonenzymatic template-directed RNA synthesis inside model protocells. *Science 342*: 1098–1100.

Alberts, B. 1998. The cell as a collection of protein machines: preparing the next generation of molecular biologists. *Cell 92*: 291–294.

Allwood, A. C. 2016. Evidence of life in the earth's oldest rocks. *Nature 537*: 500–501.

Arndt, N. T., and Nisbet, E. G. 2012. Processes on the young earth and the habitats of early life. *Annual Review of Earth and Planetary Sciences 40*: 521–549.

Ashkenazi, G., et al. 2017. Systems chemistry. *Chemical Society Reviews 46*: 2543–2554.

Baker, B. S., et al. 2001. Are complex behaviors specified by dedicated regulatory genes? reasoning from *Drosophila*. *Cell 105*: 13–24.

Balus'ka, F., and Reber A. 2019. Sentience and consciousness in single cells: how the first minds emerged in unicellular species. *Bioessays 41*: 1800229.

Becerra, A., et al. 2007. The very early stages of biological evolution and the nature of the last common ancestor of the three major cell domains. *Annual Review of Ecology, Evolution and Systematics 38*: 361–379.

Beck, W. S. 1957. *Modern science and the nature of life*. New York, Harcourt Brace.

Behe, M. 2019. *Darwin devolves*. New York, Harper One.

Betts, H. C., et al. 2018. Integrated genomic and fossil evidence illuminates life's early evolution and eukaryote origins. *Nature Ecology and Evolution 2*: 1556–1562.

Blackmore, S. 2018. The hardest problem. *Scientific American 319*: 50–53.

Blount, Z. D., et al. 2012. Genomic analysis of a key innovation in an experimental population of *Escherichia coli*. *Nature 489*: 513–518.

Blum, H. F. 1968. *Time's arrow and evolution*. Princeton, Princeton University Press.

Bonduriansky, R., and Day, T. 2009. Non-genetic inheritance and its evolutionary implications. *Annual Review of Ecology, Evolution and Systematics 40*: 103–125.

Booth, A., et al. 2016. *The modern synthesis in the light of microbial genomics. Annual Review of Microbiology 70*: 279–297.

Brasier, M. 2009. *Darwin's lost world: the hidden history of animal life*. Oxford, Oxford University Press.

Bromham, L. 2016. What is a gene for? *Biology and Philosophy 31*: 103–123.

Bruessow, H. 2009. The not so universal tree of life, or the place of viruses in the living world. *Philosophical Transactions of the Royal Society of London, Series B 364*: 2263–2274.

Brunet, T. D. P., and Doolittle, W. F. 2018. *The generality of constructive neutral evolution. Biology and Philosophy 33*: 2. 10.1007/s10539-018-9614-6.

Capra, F., and Luisi, P. L. 2014. *The systems view of life: a unifying vision*. Cambridge, Cambridge University Press.

Carrol, S. B. 2005. *Endless forms most beautiful*. New York, W.W. Norton.

Carrol, S. B. 2008. Evo-Devo and an expanding evolutionary synthesis: a genetic theory of morphological evolution. *Cell 134*: 25–36.

Cavalier Smith, T. 2010. Deep phylogeny, ancestral groups and the four ages of life. *Philosophical Transactions of the Royal Society of London, Series B 365*: 111–132.

Chandler, K. A. 2001. *The mind paradigm: a unified model of mental and physical reality*. San Jose, *Author's Choice* Press.

Coen, E. 1999. *The art of genes: how organisms make themselves.* Oxford, Oxford University Press.

Colman D. R., et al. 2017. *The deep hot biosphere: twenty years of retrospection. Proceedings of the National Academy of Sciences USA 114*: 6895–6903.

Coyne, J. A. 2009. *Why evolution is true*. Oxford, Oxford University Press.

Dacks, J. B., et al. 2016. The changing view of eukaryogenesis: fossils, cells, lineages and how they came together. *Journal of Cell Science 129*: 3695–3703.

Darwin, C. 1859. *The origin of species by natural selection*. New York, Avenel Books, reprinted 1979.

Davidson, E. H., and Erwin, D. H. 2006. Gene regulatory networks and the evolution of animal body plans. *Science 311*: 796–800.

Dawkins, R. 1996. *Climbing Mount Improbable*. New York, W.W. Norton.

Dawkins, R. 2009. *The greatest show on earth: the evidence for evolution*. New York, Free Press.

Deamer, D., and Georgiou, C. D. 2015. Hydrothermal conditions at the origin of cellular life. *Astrobiology 15*: 1091–1095.

Dennett, D. C. 1995. *Darwin's dangerous idea: evolution and the meaning of life*. New York, Simon and Schuster.

Dennett, D. C. 2017. *From bacteria to Bach and back: the evolution of minds*. New York, W.W. Norton.

Dennett, D. C. 2018. Facing up to the hard question of consciousness. *Philosophical Transactions of the Royal Society of London, Series B 373*: (1755) 20170342.

de Vries, S., and Weijers, D. 2017. Plant embryogenesis. *Current Biology 27*: R870–R873.

Difrisco, J., and Jaeger, J. 2020. *Genetic causation in complex regulatory systems: an integrative dynamic perspective. BioEssays 2020*, 42, 1900226.

Doolittle, W. F. 2013. Is junk DNA bunk? A critique of ENCODE. *Proceedings of the National Academy of Sciences USA 110*: 5294–5300.

Doolittle, W. F. 2014. Natural selection through survival alone, and the possibility of Gaia. *Biology and Philosophy 29*: 415–423.

Doolittle, W. F., et al. 2002. How big is the iceberg of which organellar genes in nuclear genomes are but the tip? *Philosophical Transactions of the Royal Society of London, Series B. 358*: 39–58.

Doolittle, W. F., and Inkpen, S. A. 2018. Processes and patterns of interaction as units of selection: an introduction to ITSNTS thinking. *Proceedings of the National Academy of Sciences USA 115*: 4006–4014.

Dulbecco, R. 1985. *The design of life*. New Haven, Yale University Press.

Eiseley, L. 1975. *All the strange hours: the excavation of a life*. New York, Scribners.

Eme, L., et al. 2017. Archaea and the origin of the eukaryotes. *Nature Reviews Microbiology 15*: 711–723.

Falkowski, P. G. 2015. *Life's engines: how microbes made the world habitable*. Princeton, Princeton University Press.

Falkowski, P. G., and Isozaki, Y. 2008. The story of O_2. *Science 322*: 540–542.

Falkowski, P. G., et al. 2008. The microbial engines that drive earth's biogeochemical cycles. *Science 320*: 1034–1038.

Forterre, P., et al. 2014. Cellular domains and viral lineages. *Trends in Microbiology* 22: 554–558.

Fox, D. 2016. What sparked the Cambrian Explosion? *Nature 530*: 268–270.

Fregnac, Y. 2017. Big data and the industrialization of neuroscience: a safe roadmap for understanding the brain? *Science 294*: 470–477.

Fry, I. 2000. *The emergence of life on earth: a historical and scientific overview.* New Brunswick, Rutgers University Press.

Fuchs, G. 2011. Alternative pathways of carbon dioxide fixation: insights into the early evolution of life? *Annual Review of Microbiology 65*: 631–658.

Gee, H. 2013. *The accidental species: misunderstandings of human evolution.* Chicago, University of Chicago Press.

Gibson, D. G., et al. 2010. Creation of a bacterial cell controlled by a chemically synthesized genome. *Science 329*: 52–56.

Gilbert, S. F., et al. 2012. A symbiotic view of life: we have never been individuals. *Quarterly Review of Biology 87*: 325–3451.

Gilmour, D., et al. 2017. From morphogens to morphogenesis and back. *Nature 541*: 311–320.

Good, B. H., et al. 2017. The dynamics of molecular evolution over 60,000 generations. *Nature 551*: 45–50.

Goodsell, D. S. 2009. *The machinery of life.* New York, Springer.

Goodwin, B. C. 1994. How the leopard changed his spots: the evolution of complexity. New York, Simon and Schuster.

Gould, S. J. 1982. Darwinism and the expansion of evolutionary theory. *Science 216*: 380–387.

Gould, S. J., and Eldredge, N. 1993. Punctuated equilibrium comes of age. *Nature 366*: 223–227.

Gross, M. 2016. How life can arise from chemistry. *Current Biology 26*: R1247–1271.

Gurdon, J. B., and Bourillot, P.-Y., 2001. Morphogen gradient interpretation. *Nature 413*: 797–803.

Harold, F. M. 1986. *The vital force: a study of bioenergetics.* New York, W.H. Freeman.

Harold, F. M. 2001. *The way of the cell: molecules, organisms and the order of life.* New York, Oxford University Press.

Harold, F. M. 2002. Force and compliance: rethinking morphogenesis in walled cells. *Fungal Genetics and Biology 37*: 271–282.

Harold, F. M. 2005. Molecules into cells: specifying spatial organization. *Microbiology and Molecular Biology Reviews 69*: 544–564.

Harold, F. M. 2014. *In search of cell history: the evolution of life's building blocks.* Chicago, University of Chicago Press.

Hazen, R. M. 2005. *Gen.e.sis.* Washington, DC, Joseph Henry.

Henikoff, S., and Greatly, J. H. 2016. *Epigenetics, cellular memory and gene regulation. Current Biology 26*: R644–R648.

Hepler, P. K. 2016. The cytoskeleton and its regulation by calcium and protons. *Plant Physiology 170*: 3–22.

Hepler, P. K., et al. 2013. Control of cell wall extensibility during pollen tube growth. *Molecular Plant 6*: 998–1017.

Higgs, P. G., and Lehman, N. 2015. The RNA world: molecular cooperation at the origins of life. *Nature Reviews Genetics 16*: 7–17.

Huang, S. 2011. The molecular and mathematical basis of Waddington's epigenetic landscape: a framework for post-Darwinian biology? *Bioessays 34*: 149–157.

Hudson, R., et al. 2020. CO_2 reduction driven by a pH gradient. *Proceedings of the National Academy of Sciences USA 117*: 22873–22879.

Hutchinson, C. A., et al. 2016. *Design and synthesis of a minimal bacterial genome. Science 357*: 1414.

Imachi, H., et al. *2020. Isolation of an archaeon at the prokaryote–eukaryote interface. Nature 577*: 519–525.

Jablonska, E., and Raz, G. 2009. Transgenerational epigenetic inheritance: prevalent mechanisms and implications. *Quarterly Review of Biology 84*: 131–176.

Jacob, F. 1973. *The logic of life: a history of heredity*. New York, Pantheon Books.

Javaux, E. J. 2019. Challenges in evidencing the earliest traces of life. *Nature 572*: 451–459.

Jiang, C., et al. 2015. Mechanisms of bacterial morphogenesis: evolutionary cell biology approaches provide new insight. *Bioessays 37*: 413–425.

Jorgensen, B. B. 2011. Deep subseafloor microbial cells on physiological standby. *Proceedings of the National Academy of Sciences USA 108*: 18193–18194.

Jorgensen, B. B., and d'Hondt, S. 2006. A starving majority deep beneath the seafloor. Science *314*: 932–934.

Josselyn S. A., et al. 2015. Finding the engram. *Annual Review of Neuroscience 16*: 521–534.

Judson, H. F. 1979. *The eighth day of creation*. New York, Simon and Schuster (Reprinted and expanded, 1996. New York, Cold Spring Harbor Press.

Kandel, E. R. 2001. The molecular biology of memory storage: a dialogue between genes and synapses. *Science 294*: 1030–1038.

Karsenti, E. 2008. Self-organization in biology: a brief history. *Nature Reviews Molecular Cell Biology 9*: 255–2672.

Kauffman, S. 2000. *Investigations*. New York, Oxford University Press.

Kauffman, S. 2019. *A world beyond physics: the emergence and evolution of life*. Oxford, Oxford University Press.

Keneally, C. 2018. Talking through time. *Scientific American 319*: 55–59.

Knoll, A. H. 2003. Life on a young planet. Princeton, Princeton University Press.

Knoll, A. H., et al. 2006. Eukaryotic organisms in Proterozoic oceans. *Philosophical Transactions of the Royal Society of London, Series B 361*: 1023–1038.

Koch, C. 2018. What is consciousness? *Nature 557*: S59–S62.

Koonin, E. V. 2015. The turbulent network dynamics of microbial evolution and the statistical tree of life. *Journal of Molecular Evolution 80*: 244–250.

Koonin, E. V. 2012. *The logic of chance: the nature and origin of biological evolution*. Upper Saddle River, FT Press Science.

Krupovic, M., et al. 2011. Genomics of bacterial and archaeal viruses: dynamics within the prokaryotic virosphere. *Microbiology and Molecular Biology Reviews 75*: 610–635.

Laland, K. 2015. How we became a different kind of animal: an evolved uniqueness. *Scientific American 319*: 322–339.

Lane, N. 2009. *Life ascending: the ten great inventions of evolution*. Oxford, Oxford University Press.

Lane, N. 2015. *The vital question: energy, evolution and the origin of complex life*. New York, W.W. Norton.

Lane, N. 2020. How energy flow shapes evolution. *Current Biology 30*: R1–R7.

Lane, N., and Martin, W. F. 2010. The energetics of genome complexity. *Nature 467*: 929–934.

Lane, N., and Martin, W. F. 2013. The origin of membrane bioenergetics. *Cell 151*: 1406–1416.

Lartigue, C., et al. 2007. *Genome transplantation in bacteria: changing one species into another. Science 317*: 632–638.

Lenton, T. M., and Latour, B. 2018. Gaia 2.0. *Science 361*: 1066–1068.

Lents, N. H., et al. 2019. The end of evolution? *Nature 360*: 590.

Lopez-Garcia, P., and Moreira, D. 2019. Eukaryogenesis, a syntrophy affair. *Nature Microbiology 4*: 1068–1070.

Lovelock, J. E. 2003. The living earth. *Nature 426*: 769–770.

Lynch, M. 2007. The frailty of adaptive hypotheses for the origin of organismal complexity. *Proceedings of the National Academy of Sciences USA 104*: 8597–8604.

Lyons, T. W., et al. 2014. The rise of oxygen in Earth's early ocean and atmosphere. *Nature 506*: 307–315.

Margolin, W. 2009. Sculpting the bacterial cell. *Current Biology 190*: R812–R822.

Margulis, L., and Sagan, D. 1995. *What is life?* Berkeley, University of California Press.

Margulis, L., and Sagan D. 2002. *Acquiring genomes: a theory of the origin of species.* New York, Basic Books.

Margulis, L., and Schwartz, K. V. 1998. *Five kingdoms: an illustrated guide to the phyla of life on earth.* San Francisco, W.H. Freeman.

Mathur, J. 2006. Local interactions shape plant cells. *Current Opinion in Cell Biology 18*: 1–7.

Maynard Smith, J. 1986. *The problems of life.* New York, Oxford University Press.

Maynard Smith, J., and Szathmary, E. 1995. *The major transitions of evolution.* Oxford, Oxford University Press.

Mayr, E. 1982. *The growth of biological thought.* Cambridge, MA, Harvard University Press.

Mayr, E. 1997. *This is biology: the science of the living world.* Cambridge, MA, Harvard University Press.

McInerney, J. O., et al. 2008. The prokaryotic tree of life: past, present . . . and future? *Trends in Ecology and Evolution 23*: 276–281.

McInerney, J. O., et al. 2014. The hybrid nature of the Eukaryota and a consilient view of life on earth. *Nature Reviews Microbiology 15*: 449–455.

McLysaght, A., and Hurst, L. D. 2016. Open questions in the study of de novo genes. *Nature Reviews Genetics 17*: 567–578.

McShea, D., and Brandon, R. N. 2010. *Biology's first law: the tendency for diversity and complexity to increase in evolutionary systems.* Chicago, University of Chicago Press.

McShea, D. W. 2015. Review of Bernd Rosslenbroich: on the origin of autonomy. *Biology and Philosophy 30*: 439–446.

Midgley, M. 1992. *Science as salvation.* London, Routledge.

Mitchell, P. 1979. David Keilin's respiratory chain and its chemiosmotic consequences. *Science 206*: 1148–1159.

Monod, J. 1971. *Chance and necessity.* New York, Random House.

Muchowska, K. B., et al. 2019. Synthesis and breakdown of universal metabolic precursors promoted by iron. *Nature 569*: 104–107.

Nagel, T. 2012. *Mind and cosmos: why the materialist neo-Darwinian conception of nature is almost certainly false.* New York, Oxford University Press.

Nicholls, D. G., and Ferguson, S. J. 1992. *Bioenergetics 2.* London, Academic Press.

Noble, D. 2017. *Dance to the tune of life: biological relativity.* Cambridge, Cambridge University Press.

Noble, D. 2006. *The music of life: biology beyond the genome*. Oxford, Oxford University Press.

O'Malley, M. A., and Powell, R. 2016. Major problems in evolutionary transitions: how a metabolic perspective can enrich our understanding of macroevolution. *Biology and Philosophy 31*: 159–189.

O'Malley, M. A. 2014. *Philosophy of microbiology*. Cambridge, Cambridge University Press.

Orgel, L. 2004. Prebiotic chemistry and the origin of the RNA world. *Critical Reviews in Biochemistry and Molecular Biology 39*: 99–103.

Pace, N. R. 2009. Mapping the tree of life: progress and prospect. *Microbiology and Molecular Biology Reviews 73*: 565–576.

Pace, N. R., et al. 2012. Phylogeny and beyond: scientific, historical and conceptual significance of the first tree of life. *Proceedings of the National Academy of Sciences USA 109*: 1011–1018.

Parfrey, L. W., and Lahr, D. J. G. 2013. Multicellularity arose several times in the evolution of eukaryotes. *Bioessays 35*: 339–347.

Pascal, R., and Pross, A. 2016. The logic of life. *Origins of Life and Evolution of the Biosphere. 46*: 507–513.

Pedrós-Alió, C., and Manrubia, S. 2016. The vast unknown microbial biosphere. *Proceedings of the National Academy of Sciences USA 113*: 6585–6587.

Pennisi, E. 2018. The power of many. *Science 360*: 1388–1391.

Pepper, J. W., and Herron, M. D. 2008. Does biology need an organism concept? *Biological Reviews of the Cambridge Philosophical Society 83*: 621–627.

Phillips, R., and Milo, R. 2009. A feeling for the numbers in biology. *Proceedings of the National Academy of Sciences USA 106*: 21465–21471.

Popper, K. R. 1972. *Objective knowledge: an evolutionary approach*. Oxford, Clarendon Press.

Pradeu, T. 2016. Organisms as biological individuals? Combining physiological and evolutionary individuality. *Biology and Philosophy 31*: 797–813.

Preiner, M., et al. 2020. A hydrogen-dependent geochemical analogue of primordial carbon and energy metabolism. *Nature Ecology and Evolution 4*: 534–542.

Pressman, A., et al. 2015. The RNA world as a model system to study the origin of life. *Current Biology 25*: R953–R963.

Pross, A. 2003. The driving force for life's emergence: kinetic and thermodynamic considerations. *Journal of Theoretical Biology 220*: 393–406.

Pross, A. 2013. The evolutionary origin of biological function and complexity. *Journal of Molecular Evolution 76*: 185–191.

Pross, A. 2016. *What is life? How chemistry becomes biology*. Oxford, Oxford University Press.

Queller, D. C., and Strassman, J. E. 2009. Beyond society: the evolution of organismality. *Philosophical Transactions of the Royal Society of London, Series B 364*: 3143–3155.

Rafelski, S. M., and Marshall, W. F. 2009. Building the cell: design principles of cellular architecture. *Nature Reviews Molecular Cell Biology 9*: 593–602.

Ramirez, S. 2018. Crystallizing a memory. *Science 360*: 1182–1183.

Reading, R. 2006. The biological nature of meaningful information. *Biological Theory 2006*: 243–249.

Richter, D., and King, N. 2013. The genomic and cellular foundations of animal origins. *Annual Review of Genetics 47*: 509–537.

Riedl, R. 1978. *Order in living organisms*. New York, Wiley.

Riquelme, M. 2013. Tip growth in filamentous fungi: a road trip to the apex. *Annual Review of Microbiology 67*: 587–609.

Riquelme, M., et al. 2018. Fungal morphogenesis, from the polarized growth of hyphae to complex reproduction and infection structures. *Microbiology and Molecular Biology Reviews*. 1–45. doi 10.1128/MMBR.00068-17.

Rogers, K. W., and Schier, A. F. 2011. Morphogen gradients: from generation to interpretation. *Annual Review of Cell and Developmental Biology 27*: 377–407.

Rosslenbroich, B. 2006. The notion of progress in evolutionary biology: the unresolved problem and an empirical suggestion. *Biology and Philosophy 21*: 41–70.

Rosslenbroich, B. 2014. *On the origin of autonomy: a new look at the major transitions in evolution*. Heidelberg, Springer.

Ruse, M. 2013. *The Gaia hypothesis: science on a pagan planet*. Chicago, University of Chicago Press.

Russell, M. J., et al. 2013. The inevitable journey to being. *Philosophical Transactions of the Royal Society of London, Series B 368*: 2012054.

Sapp, J. 2003. *Genesis: the evolution of biology*. Oxford, Oxford University Press.

Scepanski, J. T., and Joyce, G. F. 2014. A cross-chiral polymerase ribozyme. *Nature 515*: 440–442.

Schlipp, P., ed. 1949. *Albert Einstein: philosopher-scientist*. Cambridge, Cambridge University Press.

Schneider, E. D., and Sagan, D. 2005. *Into the cool: energy flow, thermodynamics and life*. Chicago, University of Chicago Press.

Schroedinger, E. 1944. *What is life?* Cambridge, Cambridge University Press.

Sebé-Pedrós, A., et al. 2017. The origin of metazoa: a unicellular perspective. *Nature Reviews Genetics 18*: 498–512.

Service, R. F. 2019. Seeing the dawn. *Science 362*: 116–119.

Shapiro, J. A. 2011. *Evolution: a view from the 21st century*. Upper Saddle River, FT Press Science.

Shapiro, J. A. 2013. How life changes itself: the read-write (RW) genome. *Physics of life reviews 10*: 287–323.

Shorto, R. 2008. *Descartes' bones*. New York, Vintage Press.

Sousa, F. L., et al., 2013. Early bioenergetic evolution. *Philosophical Transactions of the Royal Society of London, Series B 368*: 20130088.

Suddendorf, T. 2018. Inside our heads. *Scientific American 319*: 44–47.

Surovtsev, I. V., and Jacobs-Wagner, C. 2018. Subcellular organization: a critical feature of bacterial cell replication. *Cell 172*: 1271–1293.

Szathmary, E. 2015. Towards major evolutionary transition theory 2.0. *Proceedings of the National Academy of Sciences USA 12*: 10104–10111.

Taylor, A. F. 2016. Small molecular replicators go organic. *Nature 537*: 627–628.

Trevors, J. T., and Abel, D. L. 2004. *Chance and necessity do not explain the origin of life*. Cell Biology International 28: 729–739.

Turner, J. S. 2017. *Purpose and desire: what makes something "alive" and why modern Darwinism has failed to explain it*. New York, Harper One.

Waddington, C. H. 1957. *The strategy of the genes*. London, Allen and Unwin.

Ward, P., and Kirshvink, J. 2015. *A new history of life*. New York, Bloomsbury Press.

Watson, T. 2019. The trickster microbes shaking up the tree of life. *Nature 569*: 322–324.

Weiss, M. C., et al. 2016. The physiology and habitat of the last universal common ancestor. *Nature Microbiology 1*: 16116.

Wicken J. 1987. *Evolution, thermodynamics and information: extending the Darwinian program*. New York, Oxford University Press.

Woese, C. W. 2004. A new biology for a new century. *Microbiology and Molecular Biology Reviews 68*: 173–186.

Woese, C. R. 1998. The universal ancestor. *Proceedings of the National Academy of Sciences USA 95*: 6854–6859.

Woese, C. R., et al. 1990. Towards a natural system of organisms: proposal for the domains Archaea, Bacteria and Eucarya. *Proceedings of the National Academy of Sciences USA 87*: 4576–4579.

Wolpert, L. 2011. *Developmental biology: a very short introduction*. Oxford, Oxford University Press.

Wong, K. 2018. Why did *Homo sapiens* alone survive to the modern era? Last hominin standing. *Scientific American 319*: 64–69.

Zaremba- Niedzwiedzka, K., et al. 2017. Asgard archaea illuminate the origin of eukaryotic cellular complexity. *Nature 541*: 353–358.

Index

For the benefit of digital users, indexed terms that span two pages (e.g., 52–53) may, on occasion, appear on only one of those pages.

Tables and boxes are indicated by *t* and *f* following the page number.